Infrastructure Investment

An Engineering Perspective

Infrastructure Investment

An Engineering Perspective

David G. Carmichael

CRC Press
Taylor & Francis Group
Boca Raton London New York

CRC Press is an imprint of the
Taylor & Francis Group, an **informa** business

A SPON PRESS BOOK

CRC Press
Taylor & Francis Group
6000 Broken Sound Parkway NW, Suite 300
Boca Raton, FL 33487-2742

First issued in paperback 2019

© 2015 by Taylor & Francis Group, LLC
CRC Press is an imprint of Taylor & Francis Group, an Informa business

No claim to original U.S. Government works

ISBN-13: 978-1-4665-7669-8 (hbk)
ISBN-13: 978-0-367-37820-2 (pbk)

Visit the Taylor & Francis Web site at
http://www.taylorandfrancis.com

and the CRC Press Web site at
http://www.crcpress.com

Contents

Preface

For any potential infrastructure or asset, such as investments, it is necessary to establish viability, both in monetary and nonmonetary terms. *Infrastructure* and *assets* here refer to buildings, roads, bridges, dams, pipelines, railways and similar, and facilities, plants, equipment and similar. Nonmonetary issues have, for example, social, environmental and technical origins.

Future demand and future costs for infrastructure and assets generally are uncertain in both timing and magnitude. As well, other investment viability analysis input such as interest rates are similarly uncertain. The book provides the methodology by which the viability of infrastructure or any asset may be appraised as investments, given future uncertainty.

In comparison, most conventional and existing practices ignore uncertainty or attempt to include it in deterministic (that is, assuming certainty) ways, often nonrationally. For example, using interest rates adjusted for investment uncertainty is common practice, as is sensitivity analysis. Business-as-usual might be assumed to repeat in the future, even though everyone knows this is not true, but it is convenient for people to assume so, and because usable approaches incorporating true uncertainty have heretofore been unavailable. Even climate change is commonly treated deterministically in existing approaches.

The book has particular applicability for decision makers presently struggling with analyzing investments with uncertain futures, including the impact of climate change and the possible use of adaptive and flexible solutions, capable of responding to changed futures, and how such uncertainty impacts the future performance of these investments.

The book represents an original contribution to investment viability analysis under uncertainty. Existing texts have not ventured into this territory, but have rather stayed with restrictive deterministic treatments. Additionally, the book gives a very systematic and ordered treatment of its subject matter. The level of required mathematics is no more than that which is familiar to undergraduates.

The formulations given provide interesting insight into investment viability calculations and will be of use to practitioners engaged in investigatory work, especially those looking at investment risk. The material presented on options analysis opens this area to all users, breaking the confines of existing financial options analogies.

The book will be of interest to students, academics and practitioners dealing with decision making on infrastructure, assets and like investments. It will be of interest to those engaged in investments, and the analysis of real options and financial options. The content is presented in straightforward terms in order to ensure as wide a readership as possible.

The book leads the reader through a structured flow, from a systematic treatment of conventional deterministic approaches to a complete probabilistic treatment incorporating uncertainty.

About the author

David G. Carmichael is professor of civil engineering and former head of the Department of Engineering Construction and Management at the University of New South Wales, Australia. He is a graduate of the Universities of Sydney and Canterbury; a fellow of the Institution of Engineers, Australia; a member of the American Society of Civil Engineers; and a former graded arbitrator and mediator. Professor Carmichael publishes, teaches and consults widely in most aspects of project management, construction management, systems engineering and problem solving. He is known for his left-field thinking on project and risk management (*Project Management Framework*, A. A. Balkema, Rotterdam, 2004), project planning (*Project Planning, and Control*, Taylor & Francis, London, 2006), and problem solving (*Problem Solving for Engineers*, CRC Press/ Taylor & Francis, London, 2013).

Chapter 1

Introduction

1.1 APPRAISAL

The appraisal of potential infrastructure or assets, as investments, looks at the *benefits* and *costs* of everything related to the investments, both now and into the future. Common benefits include ongoing rental or product sales, and salvage or residual value at end-of-life. Common costs include initial capital cost; ongoing operation and maintenance costs; refurbishment, renovation or retrofitting costs; and disposal costs at end-of-life. *Infrastructure* and *assets* here refer to buildings, roads, bridges, dams, pipelines, railways and similar, and to facilities, plant, equipment and similar. An appraisal of a potential investment assists in

- (For a *single investment*, or for each investment) Establishing whether it is worthwhile proceeding with the investment. In other words, is the investment *viable*?
- (For *multiple investments*) Selecting between alternative investments (*preference*). In other words, which is the best investment?

Other names for *appraisal* include evaluation, study, analysis, feasibility study, benefit–cost analysis, and cost–benefit analysis. Of the last two names, engineers tend to use the second last version and stress the benefits side of the equation, much as engineers prefer to look at a glass half full, rather than half empty.

An appraisal involves consideration of issues that are

- Financial
- Nonfinancial (for example, environmental, social and technical)

That is, benefits and costs can be wider than just money. The social and environmental issues might be called *intangibles* and will have units of measurement that are not money. The units of measurement of benefits and costs need not be money, though many people find it hard or impossible thinking of units of measurement other than money. This rigidity in thinking is changing with time as environmental and social issues become more dominant in the eye of the public and start to enter people's thinking.

An appraisal can be carried out for every *stakeholder* involved in a potential investment. The data feeding into each stakeholder appraisal, of course, will be different depending on the concerns of the stakeholder. That is, it is possible for each stakeholder to come to a different conclusion as to viability and to the preferred investment. This raises serious issues of how the different viewpoints of different stakeholders are resolved, particularly in any investment that impacts on the public and on public interest (pressure) groups.

A financial appraisal would generally only involve items that can be expressed directly in money units and that affect the balance sheet and cash flow (money coming in and money going out) of the stakeholder. Typically, financial appraisals are done by the private sector. An economic appraisal, on the other hand, involves intangibles comprising environmental and social concerns, technical performance and so on, as well as the direct money items. An attempt may be made to put a monetary value on social and environmental items, but this is controversial. Typically, economic appraisals are done by the public sector. In some circumstances, a financial appraisal might be regarded as a special case of an economic appraisal. But there are multiple uses of the terminology. The mathematical manipulations for the financial and economic appraisals are the same if the intangibles are converted to money equivalents. Commonly, all benefits and costs are converted to a money unit for convenience, though the rationale behind this conversion is questioned by many people. The material in this book refers to both of what might generally be called economic appraisals and financial appraisals.

1.2 OUTLINE

This book gives the tools necessary for the appraisal of investments in infrastructure and other assets. The emphasis in this book is on dealing with uncertainty in investments (Part II: Probabilistic). However, it is considered that an introduction to deterministic analysis (Part I: Deterministic) is useful for understanding purposes. Most people learn by going from the particular to the general, and hence understanding Part I represents a necessary step in order to understand Part II.

'The term "probabilistic" implies some variability or uncertainty. "Deterministic", on the other hand, implies certainty. Deterministic variables are commonly described in terms of their mean, average or expected values. Probabilistic variables are commonly described in terms of probability distributions, or if using a second order moment approach, in terms of expected values and variances (standard deviation squared); the variances capture the variability information or uncertainty. Risk, for example, only exists in the presence of uncertainty, and hence risk approaches are probabilistic. With certainty, there is no risk. Generally, determinism is simpler to deal with and, wherever possible, people simplify from probabilistic to deterministic approaches' (Carmichael, 2013).

1.2.1 Part I: Deterministic

Part I is a stepping stone to Part II. It introduces many of the necessary terms used in investment appraisal. It discusses interest (and discount) rate matters, including compound interest for single and series amounts. A brief review of the concepts of compounding and discounting, and discounting equations is given. This leads to the various measures of appraisal, all of which rely on the time value of money. Information on investment viability and preference and the use of discount or compound interest equations in the various measures of appraisal is given. Last, Part I looks at issues to do with the choice of interest (and discount) rates, inflation, nonfinancial matters, sensitivity and benefits. A discussion of how investors treat uncertainty within a deterministic framework is given. Broadly, all these topics might be identified as belonging to what is called discounted cash flow (DCF) analysis.

1.2.2 Part II: Probabilistic

Part II extends the deterministic investment analysis of Part I to explicitly include uncertainty, and to include it in a proper way, rather than by adjusting deterministic thinking in an ad hoc way. This implies a probabilistic formulation as the most rational way forward. The material covers analysis involving probabilistic cash flows, interest rates and investment lifespans, and shows how this can also be used in the valuation of options. The favoured and adopted way of incorporating uncertainty into the analysis in this book is through a second order moment approach where variables are characterized solely in terms of their expected values and variances, thereby not requiring information on probability distributions, which in most applications would not be available.

1.2.3 Common formulation

The common probabilistic formulation given in this book covers many applications, with each application naturally specializing it in different ways. Any investment is converted to a collection of cash flows. A spreadsheet is all that is needed to perform the calculations.

Consider a general investment, with possible cash flows extending over the life, n, of the investment. Let the net cash flow, X_i, at each time period, $i = 0, 1, 2, ..., n$, be the result of a number, $k = 1, 2, ..., m$, of cash flow components, Y_{ik}, which can be both revenue and cost related. That is,

$$X_i = Y_{i1} + Y_{i2} + ... + Y_{im} \qquad (1.1)$$

where Y_{ik}, $i = 0, 1, 2, ..., n$; $k = 1, 2, ..., m$, is the cash flow in period i of component k.

Introduce a *measure* called present worth (PW), which is the present-day value of these future cash flows. As shown in Chapter 2, the present worth is the sum of the discounted X_i, $i = 0, 1, 2, ..., n$, according to,

$$PW = \sum_{i=0}^{n} \left[\frac{X_i}{(1+r)^i} \right] \qquad (1.2)$$

where r is the interest rate (expressed as a decimal, for example a rate per period of 5 % is expressed as 0.05). The term 'discounting' (and the related 'discounted') refers to converting future cash flows to their present-day value through the medium of the interest rate, which reflects the time value of money. A measure such as PW may be used to establish investment viability, and for selecting (preference) between alternative investments.

From this, information on PW, other measures such as internal rate of return (IRR), payback period (PBP) and benefit:cost ratio (BCR) can be derived. These other measures may also be used to establish investment viability, and for selecting between alternative investments. All of this is explained in detail throughout the book.

Equations (1.1) and (1.2) can be used for both the deterministic and probabilistic cases. For the probabilistic case, Y_{im}, X_i, r, n, PW, IRR, PBP and BCR become random variables. For the deterministic case, however, this formulation is perhaps more formal than it needs to be. For the deterministic case, many people find it more convenient to think in terms of benefits (B) and costs (C) rather than cash flow components (Y) or cash flows (X).

The obtained information on measures such as present worth and internal rate of return feed into the decision making regarding investments. As such, these measures might be used as objective functions, or interpreted as constraints to be satisfied in the relevant decision-making process (Carmichael, 2013). When viewed as objective functions, what is desired is that investment that maximizes or minimizes, as the case may be, the measures of PW, IRR, PBP and BCR. When viewed as constraints, what is desired is an investment that is less than or greater than, as the case may be, given values of these measures.

1.2.4 Useful reference

Understanding investment analysis is facilitated if a systems or systematic problem-solving basis is used. In this sense, recommended collateral reading is *Problem Solving for Engineers*, D. G. Carmichael, CRC Press, Taylor & Francis, 2013, ISBN 9781466570610, Cat: K16494.

Part I

Deterministic

A systems-style treatment of established deterministic investment appraisal.

Chapter 2

Benefits, costs and time value

2.1 SINGLE AND MULTIPLE INVESTMENTS

When appraising a single potential investment, what the investor is looking to establish is whether the investment is worthwhile, equivalently its *viability*. Alternatively, the terms *feasibility* or *screening* process (separating viable from nonviable investments) might be used.

When comparing multiple potential investments, the investor is looking to establish the best or preferred alternative, that is, the investment that has *preference* over others.

This chapter looks at the background to viability and preference calculations.

The appraisal of an investment is typically carried out in a systems analysis configuration (Carmichael, 2013). The cash flows are converted to measures of present worth, internal rate of return and so on, via compound interest (which is discussed later) based expressions. Broadly, the appraisal calculations might be referred to as *discounted cash flow (DCF) analysis*, when dealing with either cash flows, or benefits and costs in money units.

For the deterministic Part I, the terms *costs* and *benefits* will be used because they simplify explanations. Benefits and costs here are deterministic. However, for the probabilistic Part II, it is necessary to speak more generally in terms of cash flows. Cash flows then become random variables. It is possible to describe the deterministic case in terms of cash flows; however, this overly complicates the deterministic case.

In any investment, there will be costs. These are inputs to the investment. As a result of the investment, benefits then result (which may be positive benefits or negative disbenefits). These are outputs or outcomes of the investment. Both *viability* and *preference* can be best thought of in terms of investment inputs (costs) and investment outputs (benefits), that is, what the investor gets out of an investment with respect to what the investor puts into the investment.

2.1.1 Notation

The main notation adopted for the deterministic Part I is as follows:

i	time or period counter, $i = 0, 1,..., n$; time may be measured in any unit, for example a day, a month or a year
n	lifespan; number of interest periods
r	interest rate (or discount rate) (expressed as a decimal, for example a rate of 5% per period is expressed as 0.05)
P	principal or present value
S_n	future value; the equivalent future amount of P accruing at a rate r for n periods
A	a uniform series amount
B	benefit
C	cost
PW	present worth
AW	annual worth
FW	future worth
IRR	internal rate of return
PBP	payback period
BCR	benefit:cost ratio

2.1.2 Viability

Commonly, a potential investment is said to be viable if the outcomes of an investment exceed what is put into the investment:

Benefits > Costs

Or expressed differently,

Benefits − Costs > 0

or

$$\frac{\text{Benefits}}{\text{Costs}} > 1$$

Additional measures of viability can be given. Later these additional measures are shown to be in terms of payback periods and interest rates. (In Part II, the definition of viability gets enlarged, when the benefits and costs become random variables.)

Viability here is a constraint (Carmichael, 2013), expressed in terms of benefits needing to exceed the costs. Satisfying the constraint means that the investment is viable; not satisfying the constraint means that the investment is nonviable.

2.1.3 Preference

Where multiple potential investments exist, a preferred investment is sought. Commonly, the preference is given to the one that maximizes the outcomes of the investment compared to what is put into the investment:

Maximum (Benefits – Costs)

or

$$\text{Maximum} \left(\frac{\text{Benefits}}{\text{Costs}} \right)$$

Additional measures of preference can be given. Later these additional measures are shown to be in terms of payback periods and interest rates.

The difference (benefits minus costs), or the ratio (benefits:costs) are objective functions (Carmichael, 2013). The preferred investment maximizes the objective function.

2.1.4 Benefits and costs

The appraisal of an investment looks at the benefits and costs of everything related to the investment, both now and into the future.

Benefits and costs are looked at from the viewpoint of the relevant stakeholder or investor. Each stakeholder has a different set of benefits and costs. What may be a cost (or benefit) to one stakeholder may not be a cost (or benefit) to another. What may be a positive benefit to one stakeholder may be a negative benefit to another.

The distinction between benefits and costs is best made by regarding costs as input to the investment, while benefits (both positive benefits and negative disbenefits) are output or outcomes from the investment (Figure 2.1). That is, anything input to an investment is a cost, while anything resulting from the investment is a benefit (positive and negative).

Each stakeholder will have its own set of investment inputs and outputs. That is, a different appraisal applies for every different stakeholder.

Figure 2.1 Distinction between costs and benefits. (From Carmichael, D. G., *Problem Solving for Engineers*, Taylor & Francis/CRC Press, London, 2013.)

Typical costs (investment inputs) include the following:

- Initial invested capital; creation cost
- Design and construction costs
- Ongoing operation costs including maintenance, taxes, and insurance; money needed to run and sustain the investment
- Refurbishment, retrofitting or renovation costs during the lifespan of the investment
- Outlays, payments
- Disposal cost at the end of the life of the investment

Typical benefits (investment outputs) include both tangibles and intangibles:

- Things such as travel time or number of accidents, for a road
- Tolls and rental collected
- Salvage value or residual value, upon reaching the end of the life of the investment
- Product or income; anything arising from the investment
- Pollution, noise, loss of amenity, social disruption, loss of flora and fauna and similar (disbenefits or negative benefits)

The benefits may be gains (+) or losses (–) to the stakeholder. Negative benefits are referred to as disbenefits.

Note that a negative benefit is not the same as a cost – refer to Figure 2.1. Many texts and people get this wrong. In these texts, typically a benefit is defined as any 'gain', while a cost is defined as a 'loss'. And they get confused between costs and negative benefits, such that if something is not a 'gain', then it is classified (wrongly) as a cost. Be aware of this when reading the literature. Always refer back to Figure 2.1 for clarification as to what is a benefit and what is a cost.

2.1.5 Units of measurement

Commonly, benefits and costs are expressed in the same unit of measurement. Typically, this will be a money unit or money per something (for example money per kilolitre capacity for a water reservoir, or money per hectare of land improved for an agricultural investment), but it does not have to be (especially in the case of public sector investments). For example for agricultural investments, consumed water may be the unit of measurement. (On the other hand, appraisals for the private sector would almost certainly have a money and cash flow bias.)

Typically, benefits and costs are also translated to a common base, such as annual values (that is, per annum or abbreviated as p.a.) or present-day values, taking into account the time value of money.

Where intangibles need to be expressed in money units, shadow prices are developed; for example attempts are made at putting a monetary value on noise, aesthetics, human life, social disruption and parkland in order to bring them into the mathematical calculations. This usually involves value judgements and hence can be quite controversial. For example what is the monetary value of a human life, parkland or endangered animal?

Where the different intangibles remain expressed in their original nonmoney units or it is considered improper to express them in money units, they may take on the form of constraints (requirements that need to be satisfied), rather than entering the calculations as benefits and costs. Alternatively, these intangibles need to be traded off against the monetary side of the investment (Carmichael, 2013).

2.1.6 Time value of money

The life of investments can be long – 25, 50, 100 years or more – and a big influence on appraisal calculations comes from the fact that money has a time value. A dollar now is not the same as a dollar in the future, because money can earn interest over time.

Appraisals accordingly may take two forms:

- Appraisals that ignore the time value of money and use raw benefits and costs might be referred to as unadjusted or *nondiscounted*. They are performed using original values of benefits and costs, which occur at different times in the life of the investment.
- In appraisals that include the time value of money, the benefits and costs, which occur at different times in the life of the investment, for comparability, are translated (this is referred to as *discounted*) to the same or common time base, which is typically, but need not be, present-day money units or annual money units. The discounting is dependent on an *interest rate*, which *reflects the time value of money*.

A set of discounting equations is available for this second form. The associated mathematical analysis is referred to as *discounted cash flow (DCF) analysis*. The equations include reference to

- The duration or life of the investment
- An interest rate (Later, a related term – discount rate – is introduced.)
- Costs and benefits, and when they occur in time

Estimates for lifespan, for interest rate and for costs and benefits, in practice, may only be approximate, or may be estimated in a somewhat arbitrary manner, which makes any appraisal much more indefinite in reality than the mathematics would have people believe.

2.1.7 Cash flow

Cash flow refers to money coming in (*cash inflow*) and money going out (*cash outflow*). The *net cash flow* refers to the difference between cash inflow and cash outflow. The term *cash flow* has many uses, but in this book, it refers to the usage given here. Cash flow might be represented schematically as in Figure 2.2.

The cash outflows are costs and negative benefits (disbenefits), while the cash inflows are benefits.

An example cash flow diagram associated with a piece of infrastructure is shown in Figure 2.3.

2.1.8 Terminology uses

Be aware that terminology may be used in different ways by different people and different texts on the topic. Some examples are given here.

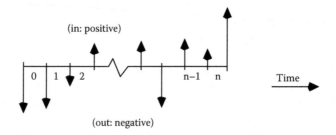

Figure 2.2 Cash flow diagram example.

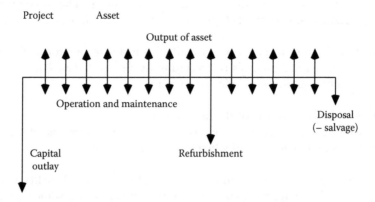

Figure 2.3 Example cash flow diagram for a piece of infrastructure; time runs from left to right.

1. Consider a project as an investment. Some people may use the term *project* to refer to all that is involved in bringing some asset into being. Others use the term to include the whole asset life cycle, or limit it to just the physical asset by itself. Many people use the term *project viability* when they mean 'project work that looks at the viability of the project and end-product of the project'. *Viability of project work* is an additional consideration, but one given less attention than *end-product viability*. In principle, the approach to looking at the viability of project work or viability of the end-product is the same; time horizons considered will be longer in the latter case, and of course the particular costs and benefits will be different.

   A project will commonly come about because of an identified need or want for some product, facility, asset, etc. This end-product is achieved through a project. However, as mentioned, the term project may also incorporate the full asset life cycle, or be the asset itself. For example a building might be a project to some people, while others might only consider a project up until the physical thing called a building comes into existence. The definition of a project is sufficiently flexible to include the operation and maintenance phase of a product within what is called the project (Carmichael, 2004).

2. The term *cash flow* may be used by some people to be equivalent to cash inflow only or cash outflow only, or the difference between cash inflow and cash outflow. Also, for example some people use this term to represent cumulative cash outflow or cumulative cash inflow. Clarification is needed as to what is intended when the term cash flow occurs in some texts and usage by people.

Confusion between these multiple usages of terminology does not appear to be a major concern to practitioners. Everything seems to work out in the end. However, be aware of multiple uses of technical terms. It would be nice to have one meaning for each word in the field of management, in order that management could progress, but this is not the reality.

2.1.9 Appraisal in stages

For large infrastructure or asset investments, appraisal might be undertaken in two or more stages. Basically, the same issues are considered in each appraisal stage, the degree of detail being least in the first stage and, as a result, involves less expenditure to undertake. Should the outcome of the first appraisal appear favourable, then the extra expense involved in the preparation of a more detailed appraisal can be justified.

There are a wide and varying number of activities, which have to be considered in a comprehensive appraisal. The order in which these activities are undertaken and their magnitude vary from investment to investment.

2.2 INTEREST

Interest reflects the time value of money; a dollar today is worth more than a dollar in the future. When money is borrowed, interest is payable to the lender. This interest may typically be calculated as a percentage of the money borrowed, and the percentage per period (usually per annum) is referred to as an *interest rate*. The interest rate represents the way money is worth more now than in the future. It allows future amounts to be converted (reduced, discounted) to present-day values.

Simple interest refers to a one-off interest payment, based on an invested amount of money (the principal). That is, at the end of the investment period, the invested amount of money gets repaid together with an interest payment based solely on the invested amount.

Compound interest refers to paying interest on interest owed, based on an invested amount of money (the principal). That is, at the end of the investment period, the invested amount of money gets repaid together with an interest payment based not only on the invested amount, but also the intermediate interest accrued.

Compounding (accruing) refers to an amount of money (the principal), subject to a given interest rate, accumulating periodically over time. That is, the invested amount grows because of interest accrued on the invested amount together with interest on the accrued interest.

Discounting refers to the relationship between the future value of an amount of money and its present value, based on assumptions about a periodic (usually annual) rate of interest and the number of compounding or interest periods (usually years). All future amounts of money can be converted to an equivalent present-day value.

In an appraisal, it is generally assumed that had money not been invested, then this money would have earned interest by being invested elsewhere. That is, if an investor uses its own money (sometimes referred to as *equity*), then interest should also be calculated on this, assuming that the money could have been invested elsewhere had it not been used in this investment. Of course, money borrowed from others (referred to as *debt*), will require interest payments at whatever interest rate is stated in the loan contract.

The following is extracted from *How to Lie with Statistics* (Huff, 1954, p. 136):

> Change-of-subject makes it difficult to compare cost when you contemplate borrowing money either directly or in the form of installment buying. Six per cent sounds like six per cent – but it may not be at all.
>
> If you borrow $100 from a bank at six per cent interest and pay it back in equal monthly installments for a year, the price you pay for

the use of the money is about $3. But another six per cent loan, on the basis sometimes called $6 on the $100, will cost you twice as much. That's the way most automobile loans are figured. It is very tricky.

The point is that you don't have the $100 for a year. By the end of six months you have paid back half of it. If you are charged $6 on the $100, or six per cent of the amount, you really pay interest at nearly twelve%.

Even worse was what happened to some careless purchasers of freezer-food plans in 1952 and 1953. They were quoted a figure of anywhere from six to twelve per cent. It sounded like interest, but it was not. It was an on-the-dollar figure and, worst of all, the time was often six months rather than a year. Now $12 on the $100 for money to be paid back regularly over half a year works out to something like forty-eight per cent real interest. It is no wonder that so many customers defaulted and so many food plans blew up.

2.2.1 Discount rate

The term *discount rate* may be preferred to be used by some people rather than *interest rate*, but mathematically they are treated the same in deterministic appraisal calculations. In the equations that follow, the symbol 'r' applies equally to interest rate and discount rate. Both interest rate and discount rate reflect the time value of money.

'Discount rate' is used to represent real change in value to the investor as determined by the possibilities for productive use of the money, and any risk associated with the use of that money. This partly implies that people should get more value from the money they borrow than the interest they pay on that borrowed money (or could receive on their own money); otherwise, there may be no point in borrowing (using) the money; that is, the discount rate chosen will be greater than the rate at which money is borrowed, or could be obtained on own funds if invested elsewhere.

Discount rates can vary from investment to investment, company to company, individual to individual. The choice of a discount rate is often a cause of disagreements in financial appraisal. One reason for this is a lack of universal agreement on how to establish its value. A second reason for this is that a discount rate is not a rational way to deal with uncertainty and risk; it is convenient to use and simplifies calculations by converting a probabilistic situation to a deterministic one, but it is nonetheless not rational.

The terms *cost of capital* and *cost of finance* and similar might be variously used by people to mean the rate applying to borrowed money. A weighted cost of capital might be used, where different weights are attached to money coming from different sources.

2.2.2 Discounting equations

In the following development, the equations and discussion make reference to an interest rate, but apply equally well to a discount rate. The discounting equations are first given for the simplest case (Section 2.2.3), namely the relationship between the value of a single amount of money now and its value in the future. This is then extended (Section 2.2.4) to give the relationship between the value of a single amount of money now and its future value spread out over the life of the investment. It is seen that the basis of all discounting equations is compound interest, manipulated to get equations into usable forms.

In a textbook treatment, particularly with textbook exercises, cash flows are usually prescribed to occur in given time periods. The question that then sometimes arises is, What happens if the cash flow occurs at the very start or very end of a period? This is only a concern in textbooks. In practice, the actual timing of any cash flow would be known, and it is that time which would be used in any calculations.

2.2.3 Single amount

Some equations that can be used to do most calculations in appraisal follow. Commonly, the time period for calculations is one year or one month, but any time period can be used. The notation of Section 2.1.1 is used.

Simple interest. An amount of money P, invested at a rate r, at the end of one period becomes,

$$S_1 = P(1 + r)$$

With simple interest calculations, interest is applied on the principal only. P might be referred to as the *principal*, but the equation is a general relationship between a future and present amount of money.

Compound interest. With compound interest calculations, interest is applied on the initial amount invested, as well as on the interest of previous periods.

An amount P accumulating at a rate r, at the end of n periods becomes,

$$S_n = P(1 + r)^n \qquad (2.1)$$

See Figure 2.4. P, again, might be referred to as the *principal*, but the equation is a general relationship between a future and present amount of money. For n = 1, the earlier simple interest equation results.

$(1 + r)^n$ is termed the *compound amount factor* (caf).

Equation (2.1) may be rewritten as

$$P = \frac{S_n}{(1+r)^n} \qquad (2.2)$$

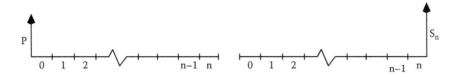

Figure 2.4 Equivalence – present value, P, and future value, S_n.

This gives the present value equivalent P to a future amount S_n occurring at period n.

$1/(1 + r)^n$ is termed the *present worth factor* (pwf).

Example

The present value of $1000 (occurring in year 5), at an interest rate of 10% per annum, assuming compound interest applies, is,

$$P = \frac{1000}{(1+0.1)^5} = \$620.92$$

2.2.4 Uniform series of amounts

For a uniform series amount, A, from Equation (2.2),

$$P = \frac{A}{(1+r)} + \frac{A}{(1+r)^2} + \ldots + \frac{A}{(1+r)^n}$$

Multiplying both sides by $(1 + r)$ gives,

$$P(1+r) = A + \frac{A}{(1+r)} + \ldots + \frac{A}{(1+r)^{n-1}}$$

Subtracting the last two expressions and rearranging gives,

$$P = A\left[\frac{(1+r)^n - 1}{r(1+r)^n}\right] \tag{2.3}$$

See Figure 2.5. This gives the present value of a uniform future series of amounts. Note, as $n \to \infty$, $P \to A/r$.

The term inside the square brackets is referred to as the *(series) present worth factor* (pwf) or present value factor.

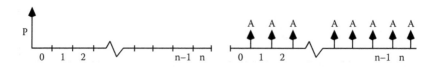

Figure 2.5 Equivalence – present value, P, and future uniform series amount, A.

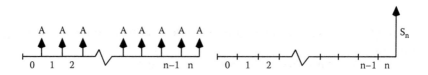

Figure 2.6 Equivalence – uniform series amount, A, and future value, S_n.

Rewriting Equation (2.3) gives,

$$A = P\left[\frac{r(1+r)^n}{(1+r)^n - 1}\right] \tag{2.4}$$

This gives the uniform future series equivalent of a present-day amount. Note, as $n \to \infty$, $A \to Pr$.

The term inside the square brackets is referred to as the *capital recovery factor* (crf).

Combining Equations (2.2) and (2.3) gives a relationship between S_n and A,

$$S_n = A\left[\frac{(1+r)^n - 1}{r}\right] \tag{2.5}$$

See Figure 2.6. This gives the future value of a uniform future series of amounts.

The term inside the square brackets is referred to as the *compound amount factor* (caf).

Rearranging Equation (2.5) gives,

$$A = S_n\left[\frac{r}{(1+r)^n - 1}\right] \tag{2.6}$$

This gives the uniform future series equivalent of a single future amount. The term inside the square brackets is referred to as the *sinking fund factor* (sff).

The above six discounting equations (Equations 2.1 to 2.6) are commonly used in appraisal. An appraisal usually involves capital expenditure, operating costs, income and so on (generally, benefits and costs), occurring at different times over the life of an investment. To obtain present values of these amounts of money, for the purpose of comparing to a common base (namely, time now, or present day), it is necessary to discount these future amounts of money using these equations. Certain additional assumptions might be made, namely that the same interest rate applies to all amounts, or different interest rates apply to cash inflows and cash outflows, and the interest rate(s) selected remains constant over the appraisal period.

2.2.5 Financial equivalence

The time value of money causes a certain present amount of money to be equal to different future amounts at different times in the future, and these change depending on the interest rate. And so it is possible to construct multiple financially equivalent scenarios, where the same present amount of money is equivalent to different future amounts, occurring at different times in the future.

2.2.6 Sensitivity of equations

The discounting and compounding equations (Equations 2.1 to 2.6) are sensitive to the interest rate, r, and the number of years, n, over which the equations are applied (as well as to any estimates made for P, A or S_n).

The present values of future amounts of money quickly become very small at nonnegligible interest rates. This means that any long-term future amounts (costs or benefits, but more usually this applies to benefits) can usually be neglected in an appraisal.

The present value of future amounts of money is sensitive to the interest rate. This sensitivity can influence preference for and against particular types of investments, depending on the future amounts involved and when in time they occur.

The present value of future amounts is sensitive to the period of appraisal selected, but more sensitive to the interest rate selected.

2.2.7 Special cases

Costs and benefits need not be characterized by a single amount or a series of uniform amounts, but might be characterized differently, for example as a uniform gradient series or geometric gradient series. It is almost certain that someone, somewhere has worked out discounting equations applying to each sort of different characterization. The so-called simulators that banks and building societies advertise to home loan customers would be particular examples. Many books have been published on the general topic. However for engineering/technical purposes, everything can be done with the above single and uniform series equations (Equations 2.1 to 2.6). The calculations might take a bit longer than using a special equation covering some different characterization, but by using a spreadsheet the calculations are relatively painless.

Exercises

2.1 How do you estimate the monetary value of a human life? What is the benefit of undertaking something that saves one human life per year? Do you calculate an average income over an average lifetime? Do you estimate average future contribution to a country's GDP? Do you calculate the average investment that the country has put into a person? Do you use insurance industry figures? Is a rich person worth more than a poor person? Is a lecturer worth more than a student? Is an educated person worth more than an uneducated one? Does your mother think that you are priceless, but everyone else thinks that you are worthless? Is there any rationality in any approach adopted?

2.2 The Rule of 72 is an old approximate method for estimating the time it will take to double an investment. Such a rule is useful because it allows for quick calculations, while being reasonably accurate. The number 72 is divisible by 2, 3, 4, 6, 8, 9 and 12 (and others), making it easy to work with manually. To calculate the time needed to double an investment, divide 72 by the interest rate. For example by investing \$1 at 8% per annum, it will take 72/8 = 9 years for the investment to be worth \$2. Alternative versions of this rule use 70 or 69 instead of 72. The rule may be extended to tripling (use 114) and quadrupling (use 144). With inflation, to determine the time for the buying power of money to halve, divide 72 by the inflation rate. For example at 3% per annum inflation, it will take 72/3 = 24 years for the value of a currency to halve. If the salary of an employee increases at a rate of 4% per annum, the salary will have doubled in 18 years (actual 17.67 years).

Examine the accuracy of the Rule of 72 for the following:
a. Interest rates per annum of 5%, 10%, 15%, 20%, 25%.
b. Using 70 or 69 instead of 72.

2.3 Plot the values obtained from the Rules of 72, 69, 70 and 76 against interest rate. Also plot the actual values. What do you conclude about these rules at different levels of interest rates up to say 20% per annum?

2.4 Develop a set of tables on a spreadsheet for interest rates of 1%, 2%, ... 25% (all per period – p.p.) in the first column, and then calculate the following in subsequent columns: (single amount) compound amount factor, present worth factor; (uniform series of amounts) sinking fund factor, capital recovery factor, compound amount factor, and present worth factor. Alternatively, set up the calculations on a spreadsheet for Equations (2.1) to (2.6).

2.5 An approximation in Equation (2.3) has $n \to \infty$, $P \to A/r$, and in Equation (2.4), has $n \to \infty$, $A \to Pr$. What is the smallest value of n for which this approximation holds? 50 years? 75 years? 100 years? More years? Test this on a range of values of n, A, P and r, and hence establish the validity of this approximation.

2.6 If you placed $3000 in your bank account on April 1 every year for the next 10 years, how much money would you have at the end (March 31) of the 11th year. Assume compounding at a rate of 10% per annum.

2.7 How much must you save annually in expenses for 10 years in order to justify an expenditure of $15 000 now (for example if you purchase a car in order to save on public transport costs)? Assume an interest rate of 5% per annum.

2.8 If you were offered a payment of $30 000 in 20 years, what payment now would be acceptable as a replacement? Use an interest rate of 5% per annum.

2.9 What replacement cost is equivalent to an annual maintenance cost of $4000 over a period of 30 years? Use an interest rate of 5% per annum.

2.10 Calculate the present value of $1 for a range of interest rates and times into the future. Use the accompanying table.

Years into the future	Present value ($) of $1 – at interest rate (p.a.) of					
	0.5%	1%	2.5%	5%	10%	25%
5						
10						
25						
50						
100						

What do you conclude from the calculations? What do the numbers imply about (a) to (d) below? Typical investments (for example in infrastructure) involve outlaying money now (costs) and recuperating benefits into the future. Or, if there are ongoing costs in the future, then the future net cash flows (benefits minus costs) are positive.

So any discussion of the impact of anything in the future typically refers to the future benefits. Consider the exercise in this light.
a. Costs and benefits into the distant future?
b. Hence, what about assumptions on lifespans?
c. Large percentage interest rates?
d. 0% per annum interest rate?

2.11 Consider a road tunnel procured by BOOT (build, own, operate, transfer) or PPP (public private partnership) delivery.

Private sector company appraisal. The private sector company involved gets to own and operate the tunnel for a period of 45 years, during which time they are able to collect tolls. At the end of the 45 years, the tunnel is handed back to the local government. The company has a large upfront cost (of the order of $2B) and maintenance and operational costs. The agreed toll amount is established in the BOOT contract and, for a car, the toll is set at $3.00 (excluding consumption taxes). Allowance is made in the contract to increase the toll according to the CPI (consumer price index),

New Toll = CPI(current year)/CPI (year contract signed) × Old Toll

Interestingly, the contract states that the toll cannot decrease even in periods of deflation.

Local government appraisal. The main costs and benefits/disbenefits associated with the tunnel are as follows:
• The toll over a 45-year period
• Some construction and maintenance/operational costs
• Disruption to the road network during construction
• Business and employment opportunities created during construction and maintenance/operation
• Additional capacity in the road network, which reduces congestion and therefore lowers travel costs and times
• Reduction in air and noise emissions to the natural environment

For the purposes of the tunnel's present worth assessment, the selected discount rate is 9.6% per annum, which incorporates a risk premium of 4.1% per annum.

The appraisal in this road tunnel example uses a discount rate of 9.6% per annum, or approximately 10% per annum. That is, benefits and costs occurring beyond about 25 years are negligible in present-day terms. How then can the private sector company justify, or the local government offer, a 45-year concession period? (The discussion here is in terms of financial justification. Do not confuse this with any benefits or not, including intangible ones, BOOT or PPP delivery may offer in terms of the private sector providing infrastructure for public use. Do not confuse this also with the fact that future costs and benefits are uncertain.)

2.12 Concession periods greater than say 30 years have in the past been justified through the introduction of an artificial device, namely a declining schedule of discount rates (the rates are assumed to reduce over time), rather than a standard fixed discount rate over the lifespan of the investment. The main purported reason for using declining discount rates is because the future is uncertain. It also gives greater present worth to money in the future, and hence makes more investments viable. The discount rate might be assumed to be declining, but is commonly approximated as a step schedule. Comment on the rationality or not of assuming declining discount rates in an appraisal.

The above description says why the device is used mathematically. But, apart from the mathematics, what is its rationality? What justification is there to say that rates actually decline with time? Why not contrive something else to get the same mathematical result, for example use a 0% per annum rate?

2.13 Calculate the present worth of $1 at the following points in time and associated interest rates:

Time (years)	Applicable interest rate (% p.a.)	Present worth of $1 ($)
5	25	
10	10	
15	5	
20	2.5	
25	1	

What do you conclude about the use of declining interest rates with time?

2.14 Accepting that climate change will lead to future obsolescence of current infrastructure, or at least the need for adaptation/modification of existing infrastructure, how might the assumption of a constant interest rate compared with an interest rate assumed to decline with time affect investment now? Consider for example future replacement costs, adaptation costs, decommissioning costs, salvage costs/value and like cash flows.

2.15 You wish to borrow money to pay for your next entrepreneurial activity. The lender offers you two repayment schemes. The going market interest rate is 10% per annum.
Scheme 1: Equal repayments of $850 each year for the next 15 years.
Scheme 2: Payments starting at $1500 in the first year and reducing by $100 each year.
Which is the better scheme?

2.16 What is the amount of money you end up with if you invest $1 at 10% per annum interest rate for 1 year, 10 years and 100 years?

2.17 What is the present worth of $1 received in 1 year, 10 years and 100 years if the interest rate is 10% per annum?

2.18 Maintenance costs are estimated at $1000 per year for 10 years. For interest rates of 5% per annum and 10% per annum, what are the present values of the cost of maintenance?

2.19 What yearly amounts are equivalent to a present value of $10 000 for an interest rate of 10% per annum over a period of 5 years?

2.20 If $100 is put aside each year at 10% per annum interest rate, what is the accumulated amount at the end of 15 years?

Chapter 3

Appraisal

3.1 MEASURES

There are a number of *measures* that find favour within appraisal for establishing the *viability* of a potential investment, and for comparing and selecting between (*preference*) alternative potential (multiple) investments:

- Present worth (PW)
- Annual worth (AW)
- Internal rate of return (IRR)
- Payback period (PBP)
- Benefit:cost ratio (BCR)

Each measure tells something different about the investment being considered, but not the whole story. The results of any appraisal have to be viewed in that light. Accordingly, people may evaluate an investment against several of these measures simultaneously (for example present worth, payback period and internal rate of return), and have in-house requirements for each measure before deciding to invest – for example, internal rate of return has to be greater than 10% per annum, and payback period has to be less than 18 months. This chapter outlines the basis of these measures.

Factors other than these measures have to be considered in any complete investment decision. These factors include the following:

- The availability of funds
- Return on investment (ROI)
- Cash flow
- Resources
- Available technology
- Stakeholders' perception
- Investment duration
- Sustainability
- Political, environmental and social factors
- Other relevant tangible and intangible factors

Also it may be better to have multiple small (and possibly diverse) investments rather than one large investment. This chapter and the following one explore some of these factors.

Some of these factors might be termed *constraints* or restrictions. For example, the available investment dollars are only $\$\alpha$, or the timeframe for the investment is to be less than β years. Generally, the presence of constraints reduces the number of potential investments (Carmichael, 2013).

3.1.1 Summary of measures

All appraisal measures are related. Where the time value of money is incorporated, all measures use the same compound interest (discounting) equations outlined in Chapter 2. All appraisal measures can be interpreted in terms of various configurations and treatments of two generic forms:

- Benefits minus costs (B − C)
- Benefits divided by costs (B/C)

Here, benefits include any disbenefits. Table 3.1 summarizes the various measures using such a framework and the notation outlined in Chapter 2.

For more than one alternative investment, alternatives might be compared pairwise and rejected one-by-one. Comparison may also be with the *status quo* or the do-nothing alternative. Where alternative investments are being considered, any matters common to the investments might be omitted.

Net cash flow is related to B − C. Return on investment (ROI) is related to B/C.

With the benefit:cost (B/C) ratio measure, care needs to be exercised because costs may be wrongly treated as negative benefits (disbenefits) and vice versa, or less commonly benefits may be wrongly treated as

Table 3.1 Summary of appraisal measures

Measure	Single investment (screening/viability)	Multiple investments (preference)
Present worth; Annual worth	B − C > 0	Max B − C
Internal rate of return	Desired r where B/C = 1 or B − C = 0	Max r where B/C = 1 or B − C = 0
Payback period (nondiscounted and discounted)	Desired n where B/C = 1 or B − C = 0	Min n where B/C = 1 or B − C = 0
Benefit:cost ratio (either in present values or annual amounts)	B/C > 1	Max B/C

negative costs; this affects the B/C ratio. Such concerns don't exist if a B – C formulation is used, or the distinction between costs and benefits is clear.

When reference is made to benefits and costs, Figure 2.1 should be kept in mind. This will overcome any doubt as to what is referred to as *disbenefits* (negative benefits). (A number of texts and people get this wrong.) Costs might be regarded by some people as positive in appraisals (but note that they are negative in cash flow diagrams and accounting practices). Benefits may be positive or negative (disbenefit). Figure 2.1 will always give correct answers for the benefit:cost (B/C) ratio measure. For the other measures, which are based on the difference B – C, it is possible to be lax in the usage as to what are benefits and what are costs, as long as the signs are properly taken care of.

3.1.2 Summary of information gained

Each measure of appraisal tells something different about the investment being considered, but not the whole story. And so it is common for practitioners to simultaneously use several measures to assist any decision making on viability or preference. As well, there is no one measure that handles all situations. Table 3.2 summarizes the information that each measure gives to the decision maker.

Table 3.2 Information gained from the different appraisal measures

Measure	Information gained
Present worth; Annual worth	The absolute difference between benefits (investment outputs) and costs (investment inputs).
Internal rate of return	The range of interest rates for which the investment is viable. The measure does not require the analyst to determine an interest rate. (The ranking of alternatives cannot be altered by the choice of interest rate.)
Payback period (nondiscounted and discounted)	The time taken for the benefits (investment outputs) to repay the costs (investment inputs).
Benefit:cost ratio (either in present values, or annual amounts)	Equivalent to an engineer's measure of productivity. Equivalently, it is an investment output/input measure, where benefits are the investment output and costs are the investment input (see Figure 2.1). A measure of investment output (benefits) per unit of investment input (costs). Related to return on investment. Relative profitability. A cost effectiveness measure allows nonmonetary units.

The information obtained from these appraisal measures is only part of the total information that feeds into full investment decision making, as mentioned at the start of this chapter.

3.1.3 Some shortcomings

Because each measure only tells part of the story and not the whole story about a potential investment, each measure of appraisal has its own shortcomings. Table 3.3 gives some of the shortcomings of each measure.

3.2 PRESENT WORTH

Other names for present worth: *present value* (PV); *net present value* (NPV).

The PW measure converts or discounts all costs and benefits into their present-day equivalent, and sums these (taking into account their signs – positive or negative). That is, the present worth measure is the difference between present-day benefits and present-day costs,

$$PW = B - C \tag{3.1}$$

When comparing alternative investments with different lifespans, the calculations are carried out over a least common multiple of these lifespans; in the analysis, the alternative with the smaller lifespan is forced

Table 3.3 Some deficiencies of the different appraisal measures

Measure	Shortcoming
Present worth	When comparing alternative investments with different lifespans, the calculations are carried out over a least common multiple of these lifespans. Possibly misleading when choosing between investments where monetary levels (investment scale) are different.
Annual worth	Possibly misleading when choosing between investments where monetary levels (investment scale) are different.
Internal rate of return	Use incremental IRR when monetary level (investment scale) is different between investments. Possibly ambiguous results.
Payback period (nondiscounted and discounted)	Susceptible to when cash flows occur. It does not acknowledge the difference between investments with different lives.
Benefit:cost ratio (either in present values, or annual amounts)	Care needed in defining disbenefits. Possibly misleading when choosing between investments where monetary levels (investment scale) are different. (Use incremental B/C ratio.)

Table 3.4 Example data

Facility	Capital cost ($)	Annual income ($)
ZI	90 000	15 000
Z2	70 000	13 000

to incorporate an amount to provide for the renewal of the investment in order that it could last the lifespan of the longer investment. The use of the same lifespans when comparing alternatives is not necessary with either *benefit:cost ratio* or annual worth measures.

Example

Two facilities are compared over a period of 25 years using an interest rate of 15% per annum. The capital cost and annual income for each are given in Table 3.4.

Incomes are first converted to their present values, and then capital costs subtracted from these. For facility Z1, the present worth measure gives,

$$-90000 + 15000\left(\frac{1.15^{25} - 1}{0.15(1.15)^{25}}\right) = \$6963$$

For facility Z2,

$$-70000 + 13000\left(\frac{1.15^{25} - 1}{0.15(1.15)^{25}}\right) = \$14034$$

Facility Z2 has the greater PW, and therefore might be considered the better investment.

PW only provides a good comparison between investments when they are strictly comparable in monetary level or total budget terms. This is because PW gives no indication of the size of investment necessary to give a particular present worth. Investment scale is not considered in the analysis.

3.2.1 Capitalized cost measure

Where the present worth calculations are carried out over an infinite time period, the term *capitalized cost*, or *capitalized equivalent*, may be used. Alternative proposals are compared based on their relative capitalized costs. The choice of the infinite time period is not liked by many because it does not reflect the actual situation in most cases.

3.2.2 Future worth

Occasionally, *future worth* (FW) might be seen in appraisal calculations. It refers to converting all benefits and costs to dollars at some nominated future point in time. In application, it is essentially the same as present worth.

3.3 ANNUAL WORTH

Other names for annual worth: *annual equivalent amount*; *equivalent uniform annual cost* (EUAC); *annual cost*; *equivalent annual worth* (EAW).

In the *annual worth* (AW) measure, all benefits and costs are converted to equivalent ongoing uniform series of annual amounts, and summed (taking into account their signs – positive or negative). That is, the annual worth (AW) measure is the annual costs subtracted from the annual benefits,

$$AW = B - C \tag{3.2}$$

The measure is related to the present worth (PW) measure (in that it also considers the difference B – C), but allows consideration of investments with different lives. PW can be converted, using the discounting equations of Chapter 2, to give annual amounts over the life of the investment.

Where benefits (and costs) are already constant from year to year, there is no difficulty with calculating AW. However, where the benefits (or costs) are irregular from year to year, it may first be necessary to convert all amounts to a single value at a common point in time. This single value is then further converted to a series of uniform amounts.

When making AW comparisons, only one lifespan of each alternative investment need be considered. AW will be the same for any number of lifespans. The AW measure is thus useful for comparing investments with different lifespans.

Example

Consider an investment scenario, with an 8% per annum interest rate:

Capital cost	$25 000
Annual cost	$3000
Salvage value	$5000
Life (years)	10

Assuming a 10-year lifespan:

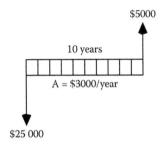

> Salvage value: PW ($5000, 10 years) = $2315
> Capital cost: PW ($25 000, 0 years) = $25 000
> Annual cost
> PW factor (r = 8% p.a., n = 10 years) = 6.7130
> PW annual cost ($3000, 10 years) = 6.7130 × 3000 = $20 139
> Total PW of costs = 25 000 + 20 139 = $45 139
> Capital recovery factor (r = 8% p.a., n = 10 years) = 0.1490
> AW (10-year period) = 0.1490 × (2315 – 45 139) = –$6380
> BCR = 2315/45 139 = 0.0513

Assuming a 30-year lifespan (with replacement every 10 years):

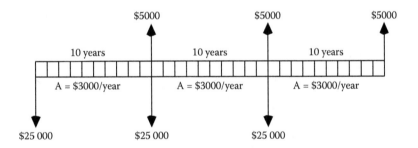

> Salvage value
> PW ($5000, 10 years) = $5000/1.08^{10} = $2315
> PW ($5000, 20 years) = $5000/1.08^{20} = $1073
> PW ($5000, 30 years) = $5000/1.08^{30} = $497
> PW salvage value = 2315 + 1073 + 497 = $3885
> Capital cost
> PW ($25 000, 0 years) = $25 000
> PW ($25 000, 10 years) = 25 000/1.08^{10} = $11 574
> PW ($25 000, 20 years) = 25 000/1.08^{20} = $5365
> PW capital cost = 25 000 + 11 574 + 5365 = $41 939

Annual cost
PW factor (r = 8% p.a., n = 30 years) = 11.2575
PW annual cost ($3000, 30 years) = 11.2575 × 3000 = $33 773
Total PW of costs = 41 939 + 33 773 = $75 712
PW (30-year period) = 3885 − 75 712 = − $71 827
Capital recovery factor (r = 8% p.a., n = 30 years) = 0.0888
AW (30-year period) = 0.0888 × (3885 − 75 712) = −$6380
BCR = 3885/75 712 = 0.0513

It is seen that, for the two different lifespans of 10 and 30 years, AW and BCR are the same, while PW is different.

Example

Data on two facilities, Z1 and Z2, are given in Table 3.5. The interest rate is 10% per annum.

For facility Z1,

Capital recovery factor (r = 10% p.a., n = 7 years) = 0.1874
AW = 2000 − 5000(0.2054) = $973/year

For facility Z2,

Capital recovery factor (r = 10% p.a., n = 5 years) = 0.2638
AW = 6350 − 20 000 (0.2638) = $1074/year

Accordingly, facility Z2 might be considered the better investment.

Example

Consider two facilities, whose details are given in Table 3.6, together with an interest rate of 10% per annum. No assumptions have been made regarding replacement.

Table 3.5 Example data

Facility	Initial cost ($)	Annual cash flow ($)	Life (years)
Z1	5000	2000	7
Z2	20000	6350	5

Table 3.6 Example data

Facility	Initial cost ($)	Annual income ($)	Design life (years)	PW ($)	AW ($)
Z1	100000	27000	5	2352	620
Z2	80000	25950	4	2259	712

Here the PW and AW measures differ as to which is the preferred alternative. For the PW and AW measures to agree, a common multiple of lives of 20 years would be necessary in the PW calculations, but not in the AW calculations.

Example

Consider the following data on a piece of equipment:

Initial cost	$30000
Anticipated life	20 years
Salvage value	$2500
Annual operating cost	$2500
Annual income	$6000
Interest rate	10% p.a.

Annual worth:

$-$ 30000 × capital recovery factor

$+$ 2500 × sinking fund factor   (salvage value)

$+$ (6000 $-$ 2500)   (annual operating cost, income)

$$= -30000 \times \frac{0.1(1.1)^{20}}{\left[(1.1)^{20} - 1\right]} + \frac{2500 \times 0.1}{\left[(1.1)^{20} - 1\right]} + 3500$$

$$= -3524 + 44 + 3500$$

$$= \$20/\text{year}$$

3.4 INTERNAL RATE OF RETURN

The *internal rate of return* (IRR) is that interest rate where the present worth is zero or the benefit:cost ratio is one. Benefits and costs may be either present values or uniform annual amounts.

IRR = r when B $-$ C = 0, or B/C = 1 $\hspace{2cm}$ (3.3)

IRR corresponds to a break-even investment, where the present value of outgoings equals the present value of any income.

The internal rate of return has an advantage over other measures. It does not require the analyst to select an interest rate. For all measures except internal rate of return, an assumption on the interest rate is required.

The internal rate of return measure, on the other hand, determines the interest rate. With other measures, the ranking of alternatives can be altered by the choice of the interest rate.

This calculation of the interest rate which is the IRR, can be programmed as a solution to an equation. Alternatively, one way of avoiding this is to perform the calculations numerically on a spreadsheet by selecting a range of interest rates and calculating the resulting present worth. That interest rate corresponding to the present worth of zero is the internal rate of return, and this may be found by interpolation.

> Under most normal circumstances, the net present value method and the internal rate of return method will give equivalent results when ranking mutually exclusive [investments]. At times, however, the two methods may give contradictory results. The conflict is primarily due to the different assumptions for the reinvestment rate of funds released from the [investments]. The internal rate of return method implies that the funds released are reinvested at the same internal rate of return over the remaining life of the proposal. The net present value method implies reinvestment at a rate equivalent to the required rate of return used as the [interest] rate.
>
> For proposals with a high internal rate of return, a high reinvestment rate is assumed: for proposals with a low internal rate of return, a low reinvestment rate is assumed. Only rarely will the internal rate of return calculated represent the relevant rate of reinvestment of intermediate cash flows.
>
> The net present value method of analysis is theoretically superior to the internal rate of return method (Antill and Farmer, 1991, p. 34).

The internal rate of return can lead to ambiguous results, because on occasions two values for the interest rate can arise. If this occurs, it is necessary to think carefully what this means.

Example

A facility costs $100 000 initially but is anticipated to return $20 000 each year over 10 years. For an interest rate of 10% per annum, Table 3.7 gives present worth calculations in column 3 and the cumulative values in column 4.

The values in column 3 of Table 3.7 are most easily obtained from Equation (2.2). The present worth at the base of column 3 could also have been obtained by using Equation (2.3).

The facility starts to produce a positive return after 7 years.

Consider some alternative interest rates. Table 3.8 and Figure 3.1 show the calculations.

Table 3.7 Example calculations

Year	Cash flow ($)	PW (10% p.a.) ($)	Cumulative PW ($)
0	-100000	-100000	-100000
1	20000	18182	-81818
2	20000	16529	-65289
3	20000	15026	-50263
4	20000	13660	-36603
5	20000	12418	-24185
6	20000	11289	-12896
7	20000	10263	-2633
8	20000	9330	6697
9	20000	8482	15179
10	20000	7711	22890
Net		22890	

Table 3.8 Example calculations

Interest rate (% p.a.)	(Series) PW factor	PW ($)
5	7.7217	54434
10	6.1446	22892
15	5.0188	376
20	4.1925	-16150
25	3.5705	-28590

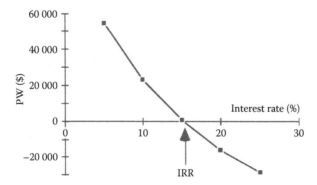

Figure 3.1 Change in PW with interest rate.

The interest rate that gives a PW of zero is approximately 15% per annum. The investment is acceptable at any rate below 15% per annum, but not above 15% per annum.

Example

A facility requires an initial investment of $100 000 and has an anticipated annual return of $30 000 for 5 years. The IRR is calculated from setting the present worth of the returns for 5 years equal to $100 000. That is,

$$30000\left[\frac{(1+r)^5-1}{r(1+r)^5}\right]=100000$$

Consider a range of values for r,

r = 10% p.a.   Left-hand side = 113 700
r = 20% p.a.   Left-hand side = 89 700
r = 15% p.a.   Left-hand side = 100 560

That is, the IRR is approximately 15% per annum.

3.4.1 Incremental rate of return

With two alternative investments of different scales, an incremental analysis, which looks at the differences between the two investments, might be used. The costs and benefits are subtracted for each year, and an IRR analysis is done on the incremental costs and benefits.

3.5 PAYBACK PERIOD

The *payback period* (PBP) measure is used to determine the period or time (usually years) required to recover an investment's outlay. The computations might be carried out in one of two ways:

- **Nondiscounted payback period** – The payback period is obtained by counting the number of years it takes for cumulative future cash flows to equal the investment outlay. Values used for B and C are nondiscounted values.
- **Discounted payback period** – The payback period is obtained by counting the number of years it takes for cumulative discounted future cash flows to equal the investment outlay. Values used for B and C are discounted values.

Table 3.9 Example PBP data

	Z1	Z2	Z3
Year 0 ($)	−2400	−2400	−2400
Year 1 ($)	600	800	500
Year 2 ($)	600	800	700
Year 3 ($)	600	800	900
Year 4 ($)	600	800	1100
Year 5 ($)	600	800	1300
Nondiscounted payback period (years)	4	3	3.3
Discounted payback period (years)	5.4	3.8	4.1
PW ($)	−125	633	867

Nondiscounted payback period is suitable for quick analyses, for analysis over short payback periods, or where the interest rate doesn't influence the calculated values significantly.

Interpolation is used between time periods to establish fractions of periods.

Example

Table 3.9, containing cash flows, indicates the payback periods and present worth for three investments, Z1, Z2 and Z3; an interest rate of 10% per annum is used.

The payback period does not acknowledge the difference between investments with different lives. It is however quite useful for comparing investments with identical lives and similar cash flows.

3.6 BENEFIT:COST RATIO

With the *benefit:cost ratio* (BCR or B/C) measure, benefits and costs can be expressed either in terms of present-day values or in terms of uniform annual series.

The measure provides a means of ranking different investments.

Care needs to be exercised when defining costs and benefits. Costs are investment inputs, and benefits (including disbenefits) are investment outputs. Ignorantly treating costs as negative benefits (disbenefits), negative benefits as costs, or benefits as negative costs, affects the BCR and will give wrong answers.

Example

Consider a road tunnel investment. If the appraisal is being done from the public's viewpoint, then the road toll paid by users is a negative benefit (disbenefit), not a cost. In using present worth (benefits minus costs), the mathematics will give the same present worth with either case. But in using the benefit:cost ratio (benefits divided by costs), the mathematics will give different answers.

3.6.1 Productivity

The benefit:cost ratio is equivalent to how engineers define productivity,

$$\text{Productivity} = \frac{\text{Output}}{\text{Input}} \qquad (3.4)$$

Here investment input (costs) and investment output (benefits) may be translated to present values or annual equivalents.

Note, this is productivity, not production. Production is similar to output, in engineering calculations.

3.6.2 Return on investment

The benefit:cost ratio is a measure similar to *return on investment* (ROI).

Example

A facility requires an initial investment of $100\,000$ and has an anticipated annual return of $30\,000$ for 5 years. What is the benefit:cost ratio, assuming an interest rate of 10% per annum?

Converting benefits and costs to present-day values,

$$\text{B/C} = \frac{30\,000\left[\dfrac{(1.1)^5 - 1}{0.1(1.1)^5}\right]}{100\,000} = 1.14$$

Example

Consider determining an appropriate passenger vessel size. Table 3.10 shows the relationship between vehicle size and estimated costs and benefits (in a money unit). Note that if a vessel was required, the 'no vessel' alternative may have negative benefits.

The results of Table 3.10 are plotted in Figure 3.2. The alternative with the greatest benefit:cost ratio is the 17.5 m vessel. Whether this alternative is the one chosen would depend on available funds and other constraints and considerations. All vessels between 12.5 m and 22.5 m are justifiable using the BCR measure (B/C ≥ 1).

Table 3.10 Benefit–cost calculations

Vessel size (m)	Costs ($)	Benefits ($)	Benefit:cost ratio
No vessel	0	0	–
10	100	75	0.75
12.5	150	155	1.03
15	190	200	1.05
17.5	240	270	1.13
20	280	290	1.04
22.5	320	320	1.00
25	370	350	0.95

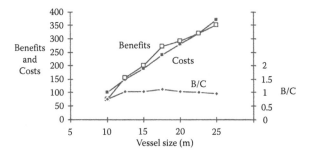

Figure 3.2 Plot of benefits and costs versus vessel size.

3.6.3 Relative profitability, incremental benefit:cost ratio

The benefit:cost ratio can be misleading when choosing between investments where monetary levels are different, because the benefit:cost ratio indicates the relative profitability of an investment. In such a situation, an incremental analysis might be attempted, looking at the differences between the two investments. (Alternatively, the present worth measure gives results in absolute terms.)

The *incremental benefit:cost ratio* does not give any indication of the viability of individual investments. The incremental benefit:cost ratio analysis might be done if the calculations of B/C and B – C imply different preferences. The incremental ratio always compares the higher cost investment against the lower cost investment. (Alternatively, subtract the smaller benefit from the larger one. A result > 1.0 suggests that the investment with the larger benefit is preferred.)

When using incremental benefit:cost ratio, the end result is a number whose sign will usually be positive when comparing viable alternatives. Regardless of whether investment Z1 is subtracted from investment Z2 or vice versa, both the numerator and denominator should have the same sign because it is likely that the investment with the higher costs will also have higher benefits. When numerator and denominator have the same sign, the result is mathematically positive.

If the sign of the ratio is negative, then it is likely that one of the investments being compared has B/C < 1.0 and B − C < 0. This investment would not be viable and would not be included in an incremental B/C analysis.

3.6.4 Net benefit:cost ratio

The benefit:cost ratio blurs the distinction between capital costs and recurring costs. This can be overcome by using the *net benefit:cost ratio*.

$$\frac{\text{Net benefits}}{\text{Investment cost}} = \frac{\text{Benefits} - \text{Recurring costs}}{\text{Initial investment}} \tag{3.5}$$

3.6.5 Cost effectiveness

Another modification of the benefit:cost ratio is the *cost effectiveness ratio*. The difference with using this measure is that benefits are expressed in physical quantities rather than monetary units. This can be particularly useful for appraising public sector investments.

$$\text{Cost effectiveness} = \frac{\text{Benefit (nonmonetary units)}}{\text{Cost}} \tag{3.6}$$

Cost effectiveness is perhaps the most general measure used in appraisal.

3.6.6 Profitability index

A further alternative to the benefit:cost ratio measure for ranking investments is the *profitability index* defined as,

$$\text{Profitability index} = \frac{\text{Net present value}}{\text{Equivalent initial investment}} \tag{3.7}$$

3.7 EXAMPLE CALCULATIONS

The following example illustrates present worth, annual worth, benefit:cost ratio, internal rate of return and payback period on the one set of data.

Consider choosing between two investment alternatives, Z1 and Z2, which have the characteristics shown in Table 3.11.

Which is the better investment? Assume a 5% per annum interest rate.

Table 3.11 Example data

	Z1	Z2
Initial cost ($)	10000	15000
Annual return ($)	2000	3000
Lifespan (years)	10	7

3.7.1 Annual worth

The calculations for AW are as follows:

Z1 (over 10 years)

AW = 2000 – 10 000(crf, 5% p.a., 10yr)

= 2000 – 10 000(0.1295) = $705

Z1 (over 20 years, assuming replication after 10 years)

First calculate, PW of 10 000 at 10 years

= 10 000(pwf, 5% p.a., 10yr)

= 10 000(0.6139) = $6139

Then, AW = 2000 – (10 000 + 6139)(crf, 5% p.a., 20yr)

= 2000 – 16 139(0.0802) = $705

Note that AW is independent of the lifespan assumed. The breakdown of the calculation for 20 years shows that AW is independent of lifespan because both pwf and crf are used and their multiplication brings the same crf (0.1295) that is derived for the smaller lifetime.

Z2 (over 7 years)

AW = 3000 – 15 000(crf, 5% p.a., 7yr)

= 3000 – 15 000(0.1728) = $408

Alternative Z1 has the greater AW, and therefore might be considered the better investment.

3.7.2 Present worth

The calculations for PW are as follows:

Z1 (over 10 years)

$$PW = -1000 + 2000(pwf, 5\% \text{ p.a.}, 10yr)$$

$$= -10\,000 + 2000(7.7217) = \$5443$$

Z1 (over 20 years, assuming replication after 10 years)

$$PW = -10\,000 - 10\,000(pwf, 5\% \text{ p.a.}, 10yr)$$

$$+ 2000(pwf, 5\% \text{ p.a.}, 20yr)$$

$$= -10\,000 - 10\,000(0.6139) + 2000(12.462) = \$8785$$

Note that PW is dependent on the lifespan assumed. This is so because both annual return and annual cost are changed with the contribution of pwf. As a result, the 10-year investment and 20-year investment have different PW values. It is not possible to compare them here. Although the 10-year investment has been duplicated once after year 10, the PW does not double.

Z2 (over 7 years)

$$PW = -15\,000 + 3000(pwf, 5\% \text{ p.a.}, 7yr)$$

$$= -15\,000 + 3000(5.7864) = \$2359$$

Z2 (over 21 years, assuming replication after 7 and 14 years)

$$PW = -15\,000 - 15\,000(pwf, 5\% \text{ p.a.}, 7yr) - 15\,000(pwf, 5\% \text{ p.a.}, 14yr)$$

$$+ 3000(pwf, 5\% \text{ p.a.}, 21yr)$$

$$= -15\,000(1 + 0.7107 + 0.5051) + 3000(12.821)$$

$$= \$5226$$

Note that PW is dependent on the lifespan assumed. To compare Z1 and Z2, this needs to be done over the same lifespan. Here 20 years is approximately 21 years. Alternative Z1 has the greater PW, and therefore might be considered the better investment.

3.7.3 Benefit:cost ratio

The calculations for BCR are as follows. Benefit:cost ratios can be compared either on an annual series or present value basis.

Z1 using annual series, 10-year lifespan

BCR = 2000/[1000(crf, 5% p.a., 10yr)] = 2000/1295 = 1.54

Z1 using annual series, 20-year lifespan (that is duplication after 10 years)

BCR = 2000/1295 = 1.54

Note that BCR is independent of the lifespan assumed using annual values.

Z2 using annual series, 7-year lifespan

BCR = 3000/2592 = 1.16

Z1 using present values, 10-year lifespan

BCR = 15 443/10 000 = 1.54

Z1 using present values, 20-year lifespan (that is duplication after 10 years)

BCR = [2000(12.462)]/[10 000 + 10000(0.6139)] = 1.54

Note that BCR is independent of the lifespan assumed using present values.

Z2 using present values, 7-year lifespan

BCR = [3000(5.7864)]/15 000 = 1.16

Z2 using present values, 21-year lifespan (that is duplication after 7 and 14 years)

BCR = [3000(12.821)]/[15 000 + 15000(0.7107) + 15000(0.5051)] = 1.16

Alternative Z1 has the greater BCR, and therefore might be considered the better investment.

3.7.4 Payback period

Payback period is the time taken to repay any outlay. Payback period calculations give this time, n, as the result.

Nondiscounted version

Alternative Z1, PBP = 10 000/2000 = 5 years

Alternative Z2, PBP = 15 000/3000 = 5 years

That is, the nondiscounted PBP measure cannot distinguish between alternatives Z1 and Z2.

Discounted version

Alternative Z1. The annual amount of $2000 is discounted year-by-year and accumulated. The present value of $2000 is 2000(pwf, 5% p.a., n yr) where n = 1, 2, 3, ... The calculations are given in Table 3.12.
It is seen that the initial outlay of $10 000 is reached at approximately 5.8 years. This is the discounted payback period.
A similar calculation can be done for alternative Z2. This will also give a discounted payback period between 5 and 6 years.
That is, the payback period measure cannot be used to distinguish between the alternatives Z1 and Z2.

3.7.5 Internal rate of return

IRR calculations determine the interest rate r corresponding to PW = 0 or BCR = 1. IRR is distinguished from the other measures, PW, AW, PBP and BCR, which all require an assumption or estimate of r, while IRR calculates r.
IRR can be found by solving a nonlinear equation, but the easier way to obtain r is by enumeration. Values of r are guessed, the corresponding PW calculated, and IRR is obtained by interpolation. Such calculations are given in Tables 3.13 and 3.14.

Table 3.12 Discounted PBP example calculations

Year	Single amount pwf	Present worth of 2000 ($)	Cumulative present worth ($)
1	0.9524	1905	1905
2	0.9070	1814	3719
3	0.8638	1727	5446
4	0.8227	1645	7091
5	0.7835	1567	8658
6	0.7462	1492	10150

Table 3.13 IRR example calculations

Guess r	Series pwf	2000 (pwf) ($)	PW = -10000 + 2000 (pwf) ($)
5	7.722	15440	5440
10	6.145	12290	2290
15	5.019	10040	40
20	4.192	8390	-1620

Table 3.14 IRR example calculations

Guess r	Series pwf	3000 (pwf) ($)	PW = -15000 + 3000 (pwf) ($)
5	5786	17360	2360
10	4.868	14600	-400
15	4.160	12480	-2520
20	3.605	10820	-4180

Table 3.15 Example summary results

Measure	Preference
AW	Z1
PW	Z1
IRR	Z1
PBP	Can't distinguish
BCR	Z1

Alternative Z1. The IRR calculations by enumeration for alternative Z1 are given in Table 3.13.

If PW (vertical axis) is plotted against r (horizontal axis), it is seen that the curve crosses the horizontal axis (where PW = 0) at just over 15% per annum. This is the IRR for alternative Z1.

Alternative Z2. Similar calculations may be done for alternative Z2 (Table 3.14).

The IRR for alternative Z2 is also just under 10% per annum.

Alternative Z1 has the greater IRR, and therefore might be considered the better investment.

3.7.6 Summary

Table 3.15 summarizes the conclusions as to what might be considered the preferred alternative, Z1 or Z2.

Note that, generally, the PW, AW, IRR, PBP and BCR measures do not have to agree with each other; each measure gives different information on the investment.

Exercises

3.1 Consider two alternative investments:

Z1 – initial outlay of $5000 and a return of $7500

Z2 – initial outlay of $20 000 and a return of $24 000

This gives:

For Z1: B/C = 1.5

For Z2: B/C = 1.2

This suggests that Z1 may be the better choice using the benefit:cost ratio measure.

(As an aside, looking at the difference between returns and outlays:

For Z1: B – C = $2500

For Z2: B – C = $4000

This suggests that Z2 may be the better choice looking at net returns.)

Looking at differences in benefits and costs for both investments, is it appropriate to calculate the incremental benefit:cost ratio as Z2 – Z1,

$$\frac{\text{Benefit difference}}{\text{Cost difference}} = \frac{24\,000 - 7500}{20000 - 5000} = \frac{16\,500}{15000} = 1.1$$

or as Z1 – Z2,

$$\frac{\text{Benefit difference}}{\text{Cost difference}} = \frac{7500 - 24000}{5000 - 20000} = \frac{-16\,500}{-15000} = 1.1$$

In the first case, the incremental benefit:cost ratio is greater than 1.0, and so it could be argued that it is worth investing in Z2. Or is this logic incorrect?

In the second case, the incremental benefit:cost ratio is greater than 1.0 and so it could be argued that it is worth investing in Z1. Or is this logic incorrect? The cost input to the 'difference investment' is negative; is this acceptable?

3.2 Consider two alternative investments:

Z1 – initial outlay of $8000 and a return of $7500

Z2 – initial outlay of $20 000 and a return of $24 000

Examine the preference for investment Z1 or investment Z2 using the incremental benefit:cost ratio measure.

3.3 A facility constructed in one year is estimated to cost $7.8 million with a life of 50 years. Earnings are estimated at $400000 per year. What interest rate will give a benefit:cost ratio of 1.0?

3.4 A facility constructed in one year is estimated to cost $7.8 million with a life of 50 years. Estimates of its earnings are made below.

What interest rate will give a benefit:cost ratio of 1.0?

If the estimates of annual returns were reduced by 20%, what would this interest rate be?

Years after completion	Annual return ($)
1 – 5	200 000
6 – 10	400 000
Thereafter	800 000

3.5 The benefits arising from the construction of a dam to provide irrigation water are estimated as follows: in the first 5 years after completion $100000 per year; in the next 5 years $200000 per year; in the next 15 years $300000 per year; in the next 50 years $400000 per year.

What is the present worth of these benefits at completion date if the interest rate is 5% per annum?

What is the equivalent uniform annual benefit over this life of 75 years at the same interest rate?

3.6 A local council plans to purchase a new garbage truck. Two models are equally acceptable. Which truck would you recommend for purchase on financial grounds?

	Model Z1	Model Z2
Purchase price ($)	50 000	60 000
Annual operating cost ($)	9000	7000

The anticipated life of each truck is 5 years with zero salvage value. The interest rate is 5% per annum.

3.7 Compare the following two facilities, 1 and 2, using present worth over a life of 25 years at interest rates of 15% per annum and 10% per annum. Comment on the difference. The capital costs and incomes for each facility are as follows:

Facility	Capital cost $	Annual income $
1	100 000	20 000
2	60 000	11 000

3.8 Three alternative schemes, Z1, Z2 and Z3, are being considered for the provision of machinery for a pumping station. For each scheme the capital cost, annual running cost and salvage value are indicated below. Determine the most attractive proposal, assuming a constant interest rate of 8% per annum, to provide a pumping facility for an indefinite number of years. Note that the anticipated life of each scheme is different.

	ZI	Z2	Z3
Capital cost ($)	25 000	50 000	35 000
Annual cost ($)	3000	2000	2500
Salvage value ($)	5000	7000	6000
Life (years)	10	29	16

The least common multiple of the lifespans is 2320 years. Is it sufficient to consider lives as approximately 10, 30 and 15 years, based on the argument that the influence (present worth) of money in later years is very small?

3.9 Consider two investments whose cash flows are as shown, together with an interest rate of 10% per annum. Investment Z1 has a lifespan of 4 years, while investment Z2 has a lifespan of 3 years.

	Year 0 ($)	Year I ($)	Year 2 ($)	Year 3 ($)	Year 4 ($)
ZI	−1000	450	450	450	450
Z2	−600	400	400	400	

Calculate the present worth and annual worth for Z1 using a 4-year lifespan and a 12-year lifespan. Calculate the present worth and annual worth of Z2 using a 3-year lifespan and a 12-year lifespan. The least common multiple of 3 years and 4 years is 12 years. The 12-year common lifespan implies 3 repeating lots of Z1, and 4 repeating lots of Z2; that is Z1 is replaced twice, and Z2 is replaced three times. What do you conclude?

3.10 Complete the following table, applying to a single investment, for a range of interest rates.

Year ($)	Cash flow ($)	PW ($)	Cumulative PW
0	0		
I	5000		
2	5000		
3	−30 000		
4	5000		
5	5000		
6	5000		

Plot PW versus interest rates. What is the internal rate of return?

3.11 As a general statement, is it possible for the present worth to be zero at two interest rates? What does such a situation imply?

Does the presence of more than one IRR value imply something about the cash flow stream changing signs?

Is an investment that can have more than one IRR more/less/equally attractive as an investment that only has one IRR?

3.12 Compute the IRR for the following data. Comment on why the answer turns out the way it does.

Initial cost: $200 000
Annual benefit: $100 000
Lifespan: 5 years
Closure cost (in year 6): $310 000

3.13 Compute the IRR for the following data. Comment on why the answer turns out the way it does.

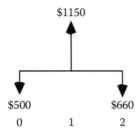

$1150

$500 $660
  0   1   2

3.14 Consider the investment cash flow of Exercise 3.13.

Write the expression for present worth in terms of the cash flows given, using an interest rate of r. This will give you a quadratic equation relating PW and r.

Plot the resulting quadratic equation in r.

Set PW = 0, and determine r.

Note that the minimum of the curve lies at r = 0.15, and the curve takes negative values between r = 0.1 and r = 0.2. How do you interpret this information?

Now consider cash flows extending in time, additional to those given in the above figure. This will generalize the above quadratic equation to something which is now a polynomial equation of the form,

$$PW = a_0 + a_1 r + a_2 r^2 + a_3 r^3 + \ldots$$

where a_j, j = 0, 1, 2, ... are constants. Such a polynomial, in theory, has multiple roots. First, are more than two roots possible? If so, what investment scenario do they represent? Second, what do all these roots mean?

3.15 Assuming that you are trying to rank two investments, each with multiple IRR values. How do you perform this ranking?

3.16 For an investment with multiple IRR values, is it counterintuitive that an investment can benefit from rising rates until a certain rate is crossed and then begin to suffer if the rate rises any further?

3.17 To compare alternatives using the AW measure, you don't need the same lifespan. AW stays the same irrespective of lifespan. But to compare alternatives using the PW measure you need the same lifespan. Why is there this difference between the PW and AW measures? What is so special about the PW measure that you need to use the same lifespan?

3.18 Comment on the following analogy that has been made by some people. Is it a good analogy or wrong?

PW is analogous to the reading on a car's odometer, which shows the total distance travelled, that is speed multiplied by time. AW is analogous to the reading on the car's speedometer, only showing the speed the car is travelling. PW is a cumulative number while AW is more of an annualized or instantaneous number. To compare PW values from two investments, it is thus necessary to 'normalize' the two investments to a common time span.

3.19 The BCR calculations give the same result using both annual series and present values, and give the same result irrespective of lifespan considered. Why should this be, when PW calculations require the same lifespan for alternatives? Why doesn't this idiosyncrasy of PW carry over to BCR when present values are used in the BCR calculations?

3.20 Draw two axes. Let the horizontal axis be the present worth of costs, and the vertical axis be the present worth of benefits. Plot the results for alternative Z2 in the example of Section 3.7 for a range of lifespans – 7, 14, 21 and 28 years. Join these four points. What does the line joining these points represent? What do you see?

From each of these four points, draw a line vertically down, of length equal to the present worth (for the respective lifespan). This will give you four more points. Now join these four new points. This line corresponds to PW = 0. What do you see?

3.21 The closer the interest rate approaches zero and the shorter the payback period, the closer the results for the nondiscounted and discounted payback periods become. As the interest rate grows and the payback period grows, so too does the difference in the payback period results. But do discounted and nondiscounted payback period versions still give the same conclusions when comparing alternatives? That is, do both measures always give that alternative Z1 is better than alternative Z2? Would you expect, in general, that the nondiscounted version of payback period will give a similar conclusion to that for discounted payback period, when comparing alternatives?

If not, can you think of an example where they would give different conclusions? The exercise is about comparing alternative investments (not comparing discounted and nondiscounted payback periods for a single investment).

3.22 The calculations in Section 3.7 for IRR were done for different lifespans for Z1 (10 years) and Z2 (7 years). Redo the calculations for IRR assuming Z1 is renewed once (20-year lifespan) and Z2 is renewed twice (21-year lifespan). Do the IRR values change? Does the conclusion change as to which is the preferred alternative via IRR?

3.23 For the example comparison in Section 3.7 between alternatives Z1 and Z2, the PBP measure could not be used to distinguish between the alternatives, yet the AW, PW, IRR and BCR measures could. Why should this be?

3.24 A benefit:cost ratio of unity corresponds to a break-even point under what conditions?

Chapter 4

Appraisal: Extensions and comments

4.1 INTRODUCTION

This chapter continues the treatment of deterministic appraisal. Additional issues related to the various measures of appraisal are discussed, including an apparent conflict between the measures, negative benefits, sensitivity and uncertainty. Comment is given on the choice of interest rate, inflation and nonfinancial matters.

4.2 APPARENT CONFLICT

Each appraisal measure (present worth, annual worth, internal rate of return, payback period and benefit:cost ratio) tells something different about the investment being considered. And the results of any appraisal have to be viewed in that light. No one measure gives the decision maker all needed information.

It is possible to get apparently conflicting conclusions among the various measures. For example, one measure might suggest investment A is preferable to investment B, while another measure might suggest otherwise.

Some situations where an apparent conflict between the measures may arise, when comparing two investments, are as follows:

- The scales of the investments are significantly different. That is, the magnitudes of the benefits and costs for one investment are much larger than for the other investment.
- The cash flow profiles of alternative potential investments are significantly different. For example, with one investment there may be cash flow in the early years, but not in later years, and this is being compared with a second investment with an opposite cash flow profile.
- The investments may have significantly different lifespans.
- The choice of interest rate determines the preferential ranking of alternatives.

Example

The conclusion from using the benefit:cost ratio measure may not agree with that using the present worth measure. Consider two alternative investments:

Z1: initial outlay of $5000 and a return of $7500
Z2: initial outlay of $20000 and a return of $24000

This gives:

For Z1: B/C = 1.5
For Z2: B/C = 1.2

This suggests that Z1 may be the better choice using the benefit:cost ratio measure.

The return on investment (ROI) for Z1 and Z2 are

For Z1: ROI = 2500/5000 = 0.5
For Z2: ROI = 4000/20000 = 0.2

This agrees with the benefit:cost ratio measure in indicating that Z1 is the better choice.

Looking at differences in benefits and costs for both investments, the incremental benefit:cost ratio becomes,

$$\frac{\text{Benefit difference}}{\text{Cost difference}} = \frac{24\,000 - 7500}{20\,000 - 5000} = \frac{16\,500}{15\,000} = 1.1$$

As the incremental benefit:cost ratio is greater than 1.0 for the additional outlay, it could be argued that it is worth investing in Z2.

Looking now at the absolute differences between the benefits and the costs,

For Z1: B − C = $2500
For Z2: B − C = $4000

This would imply that investing in Z2 is better. The B − C measure is the basis of present worth calculations and annual worth calculations.

Example

The conclusion from the present worth measure may not agree with that using the internal rate of return measure. Consider two alternative investments:

Z1: Cost $100, benefit of $150
Z2: Cost $1000, benefit of $1200

This gives:

> For Z1, PW = $50, IRR = 50% p.a.
> For Z2, PW = $200, IRR = 20% p.a.

Clearly Z1 is a better choice based on an internal rate of return measure, while Z2 is a better choice based on a present worth measure.

In such a situation, an incremental analysis might be attempted, looking at the differences between the two investments.

Between Z1 and Z2, the incremental cost is $900, and the incremental benefit is $1050, giving an internal rate of return of 17% per annum. And provided such a return is acceptable, then Z2 is the preferred alternative, and this agrees with the present worth measure conclusion.

Example

Where the present worth measure switches preference from one investment to a comparison investment dependent on the interest rate used, inconsistent conclusions may occur between the present worth measure and the internal rate of return measure.

Consider two investments, Z1 and Z2, with yearly cash flows as given in Table 4.1.

For each investment, the present worth for different interest rates, as well as the interest rate that leads to a present worth of zero, can be calculated (Table 4.2).

According to the present worth measure, Z1 is preferable for rates below 12% per annum, while Z2 is preferable above 12% per annum. However, the internal rate of return measure would always select Z2 because it gives the larger rate. Incremental internal rate of return

Table 4.1 Example cash flows

	Year 0 ($)	Year 1 ($)	Year 2 ($)	Year 3 ($)	Year 4 ($)
Z1	−1000	400	400	400	400
Z2	−500	230	230	230	230
Difference (Z1 − Z2)	−500	170	170	170	170

Table 4.2 Calculated present worth

	PW ($) (5% p.a.)	PW ($) (10% p.a.)	PW ($) (12% p.a.)	PW ($) (15% p.a.)	IRR% p.a.
Z1	409	238	174	83	18
Z2	310	211	176	123	23
Difference (Z1 − Z2)	99	27	−2	−40	12

Table 4.3 Example data and calculations

	Year 0 ($)	Year 1 ($)	Year 2 ($)	Year 3 ($)	Years 4–10 ($)	Discounted PBP (years)	PW ($)
Z1	−1000	500	500	500	0	3	243
Z2	−1000	0	0	0	500	7	829

(bottom right-hand cell of Table 4.2) indicates that the changeover rate for preferring Z2 over Z1 is 12% per annum; above 12% per annum Z2 would be preferred.

Example

For a given initial cost, but where the benefit profile over time differs between the comparison investments, the payback period measure may give a different conclusion to that of the present worth measure.

Consider two investments, Z1 and Z2, each with $1K initial cost, but with benefits as shown in Table 4.3; a 10% per annum interest rate is used.

The discounted payback period measure gives Z1 as the preferred alternative (the nondiscounted payback period measure gives the same result), while the present worth measure gives Z2 as the preferred alternative because of the greater return.

4.3 NEGATIVE BENEFITS

Generally, appraisal calculations cause no issues provided the benefits are positive. However confused thinking occurs in some texts and with some people when some of the benefits are negative, that is disbenefits exist. Once something is established as a disbenefit, there is no angst; the angst arises over what is a disbenefit. This confusion can be removed if regard is paid to Figure 2.1.

Where disbenefits are involved, it is not uncommon to see in the literature, BCR calculations where disbenefits are called costs, and thus increasing the denominator, rather than reducing the numerator. Only the latter is correct. There is only one BCR.

One suggested reason for this confusion that appears in texts and adopted by people is that layperson's meanings are being used. In layperson's terms, cost typically has a negative connotation, and benefit typically has a positive connotation and is not associated with the negative. This leads to negative benefits being referred to as costs. This situation is occurring particularly in economic appraisals because everything in such appraisals is typically reduced to a monetary value, including benefits (negative and positive). For the same lay-meaning reason, savings are also not dealt with in a consistent manner, some savings are applied as a positive benefit and some as a negative cost.

The trouble only occurs with the BCR measure. It doesn't occur with the PW, IRR or PBP measures. BCR is a ratio, as its name implies, and so what goes in the numerator and what goes in the denominator is important. PW is a difference B – C, and so people can be ill-disciplined and call a cost a negative benefit, or call a benefit a negative cost; both lead to the same answer in the deterministic case (Part I of this book) (but not the probabilistic case – Part II of this book).

4.4 UNCERTAINTY

The conclusion as to whether a potential investment is viable, or one investment is better than another, is dependent on estimated values for the following:

- The various cash inflows and outflows (capital cost, insurances, taxes, labour, materials, plant, fuel, rentals, maintenance, repairs, supervision, market prices and demand, salvage value, etc.)
- The interest rate
- The life of the investment

All of these contain uncertainty. There is always uncertainty in any appraisal. Uncertainty arises from various sources including insufficient data, having to guess the future, and dealing with intangibles, technical considerations and matters unable to be reduced to exact dollars. For example, investments may be analyzed for say 20, 25 or more years ahead, and this is based on estimates of future demand, population growth, etc.

It is remarked that the accuracy with which costs and benefits can be predicted even 10 years into the future, especially in a competitive market, is poor. Attention needs to be directed at the appraisal assumptions, but without necessarily going into meticulous detail. Appraisals might only be carried out to an order of accuracy of ±15%. Predicting competitive threats can be difficult (because of the vast array of pricing strategies available), and for that reason may be left out of any analysis. With income from mining, the uncertainty arises out of market prices for the ore, ore demand, ore grade and recoveries.

The exposure to the uncertainty in the cash flows, interest rate and lifespan is the risk.

It can be seen that an appraisal, while appearing exact because of the mathematical equations used, is not necessarily so. The conclusions can be manipulated to a certain extent.

Common practice is to assume determinism and ignore any uncertainty. This facilitates the calculations, while not requiring any advanced knowledge on probability. But there are obvious shortcomings to such an approach. Should the reality be different to any of the assumed-deterministic estimates, then the viability conclusion may change.

A number of practices are adopted to try to deal with uncertainty:

- Sensitivity analysis
- Adjusting interest rates and cash flows
- Monte Carlo simulation
- Cut-off period
- Expected values of scenarios
- Probabilistic benefit–cost analysis
- 'What if' analysis, including an examination of worst case and best case estimate scenarios

The results of these practices feed into any risk management exercise associated with the investment.

Cut-off period. Uncertainty can be crudely dealt with by imposing a cut-off period. The period is selected to ensure that the investor recovers all capital outlays. For investments containing high uncertainty, the period is chosen to be shorter than for an investment containing low uncertainty.

Expected values of scenarios. If probability estimates are assigned to alternative outcomes, expected values can be computed. For example, if an investment has two possible outcomes (scenarios), say $30 and $300 and their respective probabilities are 0.2 and 0.8, the expected value of this would be $30 \times 0.2 + $300 \times 0.8 = 246. The main issue with this method is in obtaining the estimates of the probabilities if there are no data. Educated guesses may have to be used.

4.4.1 Sensitivity analysis

To supplement the common and convenient practice of assuming determinism (even though uncertainty exists), a sensitivity analysis may be incorporated to partly acknowledge the uncertainty.

Depending on the sensitivity to the estimates of the viability conclusion, then so a decision on how to deal with the associated risk might be made.

A sensitivity analysis, in general terms, perturbs inputs (one at a time), and examines the corresponding outputs. Where a small perturbation to an input only produces a small perturbation to the outputs, the outputs are said to be *insensitive* to changes in the input. Where a small perturbation to an input produces a large perturbation to the outputs, the outputs are said to be *sensitive* to changes in the input (Carmichael, 2013). Here, analysis inputs are the values assumed (for cash flows, interest rate and lifespan),

Figure 4.1 Analysis inputs and outputs. (From Carmichael, D. G., *Problem Solving for Engineers*, Taylor & Francis/CRC Press, London, 2013.)

while analysis outputs are the appraisal measures PW, AW, IRR, PBP and BCR. See Figure 4.1.

It is generally considered good practice to carry out a sensitivity study to establish which changes in estimated values for the analysis inputs affect the conclusion (based on analysis outputs). All estimated values (for the analysis inputs) but one are held constant; only one is varied at a time, by ±x%. The process is repeated through all the analysis input variables. Such analysis can establish the point at which two alternatives are equivalent.

Criticism of assuming determinism coupled with a sensitivity analysis is that this approach is unable to judge the degree to which an investment satisfies the requirement of B – C > 0 or B/C > 1 for viability. The approach does not realistically consider the dispersion and uncertainty in, and utilize all available information about, the analysis input variables. Sensitivity analysis tries to deal with the uncertainty in viability measures by considering finite changes in the analysis input variables, but does not recognize the degree of change, or uncertainty, that the analysis input variable is actually exposed to, or produce any form of likelihood of such change. The approach is unable to differentiate grades of separation from a viability requirement; it is unable to establish the likelihoods of occurrence associated with the increments adopted in the sensitivity analysis; the probability with which the values ±x% occur is unknown and conceivably could even be zero. Part II of this book addresses this shortcoming.

For further comment on sensitivity analysis, see Carmichael (2013).

4.4.2 Adjusting interest rates and cash flows

Selecting an interest rate adjusted for risk (or discount rate) in order to allow for uncertainty compounds that uncertainty over time, and generally this doesn't reflect reality. As an alternative to the use of rates adjusted for risk, some writers suggest adjusting future benefits and costs for uncertainty. Both approaches are not rational, but rather devices used to simplify the calculations.

A more rational approach is to incorporate uncertainty in future benefits and costs, interest rate and investment lifespan in a probabilistic way; such an approach does not compound the uncertainty effect, but rather models uncertainty in a more realistic way. This is the subject of Part II of this book.

'Loading the interest rate' may also consist of reducing the rate in the case where costs are considered to have been underestimated, and increasing the rate where benefits are considered to have been overestimated. This method of incorporating uncertainty is supported by views that the profitability of an investment is more certain earlier in the life of the investment. When rates are increased to include uncertainties, early cash flows are emphasized. The approach has its critics, particularly because compounding skews the results.

4.4.3 Monte Carlo simulation

Monte Carlo simulation is a generic tool for conducting a numerical input-output analysis. It gives statistical information on the analysis outputs based on sampling probabilistic information on the analysis inputs, and a known input-output relationship (model). In general investment analysis, the analysis inputs are benefits, costs, interest rate, and lifespans, the analysis outputs are PW, IRR, PBP, BCR,..., while the input–output relationship is one of the discounting equations of Chapter 2. This is compared with deterministic benefit–cost analysis, which provides single figure analysis output.

Simulation calculations are done by computer. The same calculations are done over and over again, each time selecting different analysis input values based on sampling the distributions describing the analysis inputs. In effect, simulation converts one probabilistic analysis into multiple, repeated simple deterministic analyses. The exact forms of the distributions describing the analysis inputs are usually not known because of a lack of data, and so distributions have to be assumed.

A major advantage of simulation is that it can consider a number of uncertain analysis input variables simultaneously, choosing values according to the ranges and probabilities of each. This enables a reasonably realistic analysis. Sensitivity analysis is limited because it can handle only one or two analysis input variables at a time while holding all the others constant. Simulation goes beyond this limitation by allowing all of the analysis input variables to change at the same time.

Monte Carlo simulation will give the same results as those obtained using the book's Part II second order moment approach. The distinction is that Monte Carlo simulation is numerically based, and hence makes it difficult to draw generalizations and understanding from. A second order moment approach shows how each analysis input variable affects analysis outcomes, through taking closed-form results up to the last line of the analysis.

4.4.4 Probabilistic benefit–cost analysis

If assumptions are made on the probability characteristics describing the benefits and the costs, and the interest rate and lifespan, a probabilistic analysis may be undertaken. Working with probability distributions for the input variables leads to intractable mathematics; however, using the moments of the variables does not. The approach using moments is covered in Part II of this book.

Options analysis. Options analysis deals with uncertainty where flexible future choices are available. Part II of this book develops options analysis.

Decision making in stages can be put into an options form. For example, a utility company is uncertain about the future growth of its electrical load; and so instead of buying one large generator straightaway, it may purchase smaller generators over time if demand increases.

Instead of undertaking something straightaway, sometimes by waiting it is possible to resolve uncertainty. If events turn out to be unfavourable, the value of an initial investment may be totally lost, whereas the cost of waiting may be only the income or savings foregone until a decision is made.

4.5 CHOICE OF INTEREST RATE

The results of an appraisal can be very sensitive to the choice of interest rate. A slight change in a rate may alter the conclusion as to which is the preferred alternative investment, or can greatly affect the financial viability of an investment. The rate is also crucial in that it often determines the type of investment that is undertaken. Generally, higher rates are unfavourable to investments with initial cash outflows (capital intensive) and long-term cash inflows; the future cash flows get progressively discounted with time. There may be a choice between the alternative, which costs less initially but costs more to operate, and the alternative which involves investing more at the beginning to save on operating and maintenance costs downstream. The choice of interest rate may affect the decision as to which is more preferable.

The fact that the interest rate is a matter for judgement and can have significant financial and technical (in terms of selection of investment type) implications, can make its choice controversial.

4.5.1 Trends

Considering a general investment involving cash outflow followed by cash inflow, *decreasing the interest rate* has the following effects:

- It increases the number of investments that are viable. If, for example, the public sector uses lower rates than the private sector, then public sector activity will be more prominent.

- It favours/encourages high initial capital-intensive investments, that is where most outlay occurs early.
- It favours/encourages investments with long-term benefits, and discourages short-term investments with quick/early benefits. A high interest rate discounts significantly long-term benefits with respect to initial costs.
- Related to this last point, is that it encourages early investment activity, and similarly discourages staged investment activity. This may tie up capital and reduce future flexibility.

4.5.2 Interest rates and the long-term future

Consider Table 4.4 containing present worth factors.

It is seen that as the interest rate increases and time increases, the present value of $1 reduces to something very small.

This raises difficult ethical questions regarding long-term equity. Is it proper for people now in their decision making to ignore large costs imposed on future generations for short-term benefits now, or costs now that would result in large future benefits? Should each generation in the future be given equal weight in the analysis?

How far into the future should people attempt to carry the analysis in the evaluation of sustainability? High interest rates ignore sustainability issues. Only very low interest rates (0% to about 2% per annum) can attempt to incorporate sustainability ideas. For common commercially used interest rates, anything beyond about 25 to 50 years has negligible present worth, and these short time periods (short in the sense of mankind's time on earth) would appear to exclude sustainability.

Following are some suggested ways of incorporating these ideas:

- Adjust the interest rate to give greater weight to the future, for example use a zero interest rate (equivalent to adopting a no resource or environment depletion – strategy).

Table 4.4 Present worth factors

Years into the future	Present value of $1 at interest rate (p.a.)					
	0%	1%	2.5%	5%	10%	25%
5	1.00	0.95	0.88	0.78	0.62	0.33
10	1.00	0.91	0.78	0.61	0.39	0.11
25	1.00	0.78	0.54	0.30	0.09	—
50	1.00	0.61	0.29	0.09	0.01	—
100	1.00	0.37	0.09	0.01	—	—

- Use an interest rate based on the explicit inclusion of environmental benefits and disbenefits and application of a sustainability constraint. The sustainability constraint in its strongest form would prevent any investment proceeding that would cause a net loss to the environment.

4.5.3 Private sector work

Commonly, private sector organizations set a corporate rate, which might include the influence of the following factors:

- The cost of obtaining investment capital, which depends on
  - The general level of borrowing rates
  - The creditworthiness of the company
  - The risk associated with the investment
  - The method used to raise capital
- The opportunity cost of capital
- The return sought by the company
- The risk associated with the investment as perceived by the organization

Generally,

Adopted (discount) rate
= Basic rate + Something allowing for the productive use
of money + Something allowing for uncertainty or risk

Opportunity cost reflects the idea that had money not been put into the investment, then the money could have been employed elsewhere. The investment should return at least or better than this alternative use of the money. A minimum acceptable rate of return might reflect an organization's expectations and the interest rate charged on borrowed funds, together with an allowance for any uncertainty or risk. Commonly, investments are financed through a combination of equity and borrowed funds; a weighted rate would appear sensible in such cases, to reflect the cost of each. Revenue generated by an investment may be interpreted as equity or reduced borrowings.

With compounding, the use of a risk premium within the adopted rate means that the absolute value of the risk premium in dollars increases as time goes on, and this does not make sense for many investments that are uncertain for an initial period but tend to settle down to a known pattern of costs and revenues and therefore lower risk (real estate development, for example). This obscures the outcome of the analysis. It may be difficult to calculate the risk premium in a specific case, particularly where historical data are unavailable. Generally, interest-rate adjustment is not a rational way to deal with uncertainty.

4.5.4 Public sector work

Public sector rate selection has the added complication of public involvement, and nonfinancial benefits to be considered. Some possible methods for determining a public sector rate include the following:

- Using a zero (or very low, allowing for administration costs) rate when investments are funded from tax receipts or consideration needs to be given to the long-term future.
- A rate that reflects society's time preference rate.
- The cost of borrowed capital to the government.
- The opportunity cost of investments forgone by the private sector as a result of paying taxes.
- The opportunity cost of public sector investments foregone due to budget constraints.

What society or the public sector might like as a suitable return could be anticipated to be lower than private sector expectations because of the presence of taxation and the need for profits applying to the private sector. So-called state-owned enterprises which have an obligation to the government to make a profit while also serving the community, are a halfway house between the private sector and the public sector. This affects the choice of rate and also the way intangibles are treated.

The public sector can generally borrow at lower preferential rates than the private sector. The risk associated with the loan tends to be less, and the period of the loans longer.

Overall, the rates for public sector investment analysis could be anticipated to be lower than for private sector analysis. This will favour public sector investments over private sector investments. A zero or low rate gives an undue bias to public sector work over private sector work. Against this, public sector investments often are unattractive to the private sector, and also have some social or environmental reason.

4.6 INFLATION

Unless cash inflows and cash outflows are inflating at different rates, commonly inflation is left out of appraisal calculations.

Inflation is the term used to describe the general increase in the level of prices in the economy over time. The effect of inflation is that one dollar this year will buy less than one dollar next year.

Price indices compiled by governments are used to measure inflation. They measure the change in price of a range of commodities. The indices can be used to calculate the inflation rate.

$$\text{Annual inflation rate year}(t+1) = \frac{\text{index}(t+1) - \text{index}(t)}{\text{index}(t)} \qquad (4.1)$$

Inflation is different from the time value of money. It has an effect on the return from investments. For example, if money is invested at 10% per annum, when inflation is 15% per annum, then this is a money-losing situation.

To consider the effects of inflation in an appraisal it is necessary to differentiate between:

Nominal interest rate (r_N) The interest rate available in the market for investment or loan (sometimes called market rate).

Real interest rate (r_R) The interest rate after the effect of inflation has been taken into account.

They are related through,

$$r_R = \left(\frac{1+r_N}{1+f}\right) - 1 \qquad (4.2)$$

where f is the inflation rate.

Example

If the market interest rate is 10% per annum and inflation is 2% per annum, the real interest rate is,

$$r_R = \left(\frac{1+0.1}{1+0.02}\right) - 1 = 7.8\% \text{ p.a.}$$

Comparing this value when r_N is 18% per annum, and inflation is 8% per annum,

$$r_R = \left(\frac{1+0.18}{1.08}\right) - 1 = 9.25\% \text{ p.a.}$$

4.6.1 The effects of inflation

Cash flows can be represented in either actual dollars or constant worth dollars.

Actual dollars   Represent the current dollar values at any time during the analysis (inflation included). Sometimes called current dollars.

Constant dollars   Are dollars that have the same purchasing power at a defined point of time, for example, 2014 dollars.

They are related through,

$$\text{Constant dollars} = \frac{\text{Actual dollars}}{(1+f)^t}$$
(4.3)

where t is the time between the nominated constant dollar year and the actual dollar year.

Two methods of analysis are available:

- Using actual dollars and the nominal interest rate.
- Using constant worth dollars and the real interest rate.

4.7 NONFINANCIAL MATTERS/INTANGIBLES

In any appraisal, considerations other than financial ones can play an important role:

- Technical, including constructability and function
- Legal, including approvals
- Environmental, both statutory and community viewpoints
- Social impacts on the community
- Sustainability, and its effect on future generations
- Political, at all levels of government

Too often, appraisals concentrate on the financial side and ignore other issues. The analogy with the story about a person's night-time search for lost keys may be drawn; here the person looks for the lost keys under a lamp post, not because the keys were lost there, but because that is where the light is. Financial issues are quantifiable, and capable of being analysed; nonfinancial issues are everything else.

These issues (along with the various financial measures of PW, IRR, PBP and BCR) might be called objectives or criteria (Carmichael, 2013). There is always the dilemma of mixing quantitative and nonquantitative objectives/criteria, objectives/criteria measured in different units (*noncommensurate*) and objectives/criteria that are often *conflicting*.

4.7.1 Multiobjective/multicriteria situations

A number of approaches have been proposed for dealing with the multiobjective or multicriteria situation. Unlike the single objective/criterion case, the multiobjective/multicriteria case requires some subjective information, and hence any conclusion or outcome can always be criticized on this ground.

The multiobjective/multicriteria case can be approached in three ways (Carmichael, 2013):

- A single composite objective/criterion is made by joining all the objectives/criteria together, commonly through the choice of weightings for the objectives/criteria.
- One objective/criterion is chosen as the sole objective/criterion and the others are converted to constraints.
- The best outcome is obtained for each objective/criterion in isolation, and the resulting set of outcomes are then traded off against each other.

In the first approach the subjectivity enters through the choice of weights; in the second approach through the selection of one objective/criterion taking prominence over the others and the choice of constraint levels; in the third approach in the trading-off process.

Different people give different objectives/criteria different priorities. Combining the priorities of people introduces more subjectivity into the process.

4.7.1.1 Carbon emissions

Historically, investors considered the single objective/criterion situation of minimizing cost, or maximizing profit or return on investment. This often equated to selecting the investment that maximized the measures of PW, AW, FW, IRR and BCR, or minimized the PBP measure. To do this, all intangibles are converted to dollars; alternatively intangibles are constrained to upper or lower limits.

With carbon emissions, two approaches have been adopted by different countries:

- Carbon trading converts carbon emissions to dollars. That is, two objectives/criteria of cost/profit and emissions are converted to one objective/criterion of dollars. Trading carbon credits enables companies to sell credits as they reduce their carbon emissions, and buy them if their emissions exceed their allocated quota. The price at which these credits are traded is determined by market forces and this can be used to place a monetary value on the environmental intangible of carbon emissions.
- Regulations to force organizations to reduce emissions convert emissions to a constraint. That is, two objectives/criteria of cost/profit and emissions are converted to one objective/criterion of dollars, and one constraint of emissions. Legislation forces organizations to reduce carbon emissions, and with penalties attached, carbon emissions effectively get converted to dollars.

4.7.2 Valuing intangibles

Attempts can be made to develop *shadow prices* for those items (intangibles) that do not have a money value. A shadow price is a price for a resource, good, or service which is not based on actual market exchanges, but is derived from indirect data obtained from related markets. (Other uses of the term *shadow price* exist.)

Example

Tunnel projects may require the acquisition of a number of surface properties as well as the acquisition of subsurface land along the length of the tunnel. Compensation in law is provided, and may cover the following:

- The agreed market value of the property (a court will determine a value if agreement can't be reached)
- Costs associated with finding a replacement property
- Any charges such as legal fees, mortgage costs, valuation costs, and stamp duty payable
- Relocation costs (furniture removal, reconnection of services)

It is possible to estimate a monetary value for all these things. The law may not provide compensation for the social (or emotional, or *The Castle*, Village Roadshow, 1997) cost that a person may associate with his/her dwelling or business.

Because the social cost of resumptions is not included in a benefit–cost analysis, practices might be adopted to mitigate the resumption impacts. For example, a community consultation process may be put in place to assist residents. Urban regeneration initiatives (that is, new redevelopment sites, local amenity, increased access and improved visual character) may also be included in an attempt to mitigate the removal of businesses, residences and facilities from communities.

Maybe the true test in the public sector is that the politicians have to weigh up whether inconveniencing a few is an acceptable outcome such that many may benefit.

Example

On greenfield sites, one method used to value natural habitat is the habitat hectares method. This involves estimation of the quality and quantity of native vegetation to be removed. This is then used to calculate an equivalent offset (area to be replanted with similar vegetation). The cost of determining the offset (initial surveys, administration costs, permit costs, etc.), replanting, managing the offset and managing community expectations can be used to determine the cost of removal of the habitat.

4.7.2.1 Carbon

As mentioned earlier, trading carbon credits enables companies to sell credits as they reduce their carbon emissions, and buy them if their emissions exceed their allocated quota. The price at which these credits are traded is determined by market forces, and this can be used to place a monetary value on the environmental intangible of carbon emissions.

Some organizations are carbon neutral (through voluntary reduction) and place an internal value (or 'price') on carbon. This price is referred to when making decisions that may influence the amount of resources, for example water, energy and materials, which the organizations consume.

4.7.2.2 Nonfinancial reporting

Triple bottom line (TBL) reporting considers three impacts – financial, social and environmental. Related initiatives include corporate social responsibility (CSR) reporting, ESG (environmental, social, governance) reporting and sustainability reporting.

All recognize that it is not always appropriate to convert certain company practices into monetary units of measurement and base reporting purely on financial performance. They enable a company to report on financial, social and environmental matters, and utilize qualitative analysis and monetary and nonmonetary quantitative analysis to present a more accurate picture of a company's practices. Examples of nonmonetary quantitative analysis would be carbon footprints, solid waste volumes, health and safety statistics, and employee and customer satisfaction survey results. Qualitative analysis examples would be compliance with regulations and guidelines, commitments to sustainable development and social responsibility.

4.7.3 Equity

When costs and benefits from an investment are aggregated, as is commonly the case, this can ignore equity principles, whereby some groups or individuals may bear more/less of the costs and/or benefits than other groups or individuals. For example a road from A to B may benefit people at A and B but harm the people between A and B; it may benefit those with cars at the expense of the whole population.

Exercises

4.1 Discuss the implications of the first example in Section 4.2.

4.2 Which of the following two investments (Z1 and Z2) is better from a PW point of view, and which is better from a BCR point of view? Use a 5% per annum interest rate for the calculations. Comment on why the comparison turns out the way it does.

    Z1

        Initial cost = \$1 000 000

        Annual benefit = \$232 000

        Life = 5 years

Z2
>   Initial cost = $1000
>   Annual benefit = $400
>   Life = 5 years

Now consider an incremental analysis (Z1 – Z2).
>   Initial cost = $999 000
>   Annual benefit = $228 000
>   Life = 5 years

What do incremental PW and incremental BCR measures tell you as to which is the better investment?

Based on this example, what don't present worth and benefit:cost ratio tell you in a comparison between investments?

4.3 Consider present worth, benefit:cost ratio and incremental benefit:cost ratio measures to establish which is the preferred investment for the following data.

Z1: Benefit = 1.2; Cost = 0.5
Z2: Benefit = 2; Cost = 1

4.4 Consider two investments whose costs and benefits over a 4-year period are as shown.

	Cost ($M)	Benefit ($M)	B/C	B – C
ZI	I	1.6		
Z2	0.5	0.92		
Difference (ZI – Z2)				

Complete this table. Which is the preferred investment?

4.5 Which of the following two cash flow scenarios (Z1 and Z2) is better from a PW point of view, and which is better from an IRR point of view? Use a 10% per annum interest rate for the PW calculations. Comment on why the comparison turns out the way it does. Is the slope of the curve – PW versus r – as it crosses the r axis positive or negative in each case Z1 and Z2, and what does the sign of the slope mean in terms of interpreting IRR?

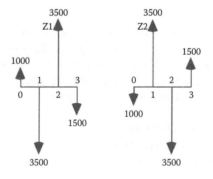

4.6 Which of the following two cash flow scenarios (Z1 and Z2) is better from a PW point of view, and which is better from a PBP (nondiscounted and discounted) point of view? Use a 10% per annum interest rate for the calculations. Comment on why the comparison turns out the way it does.

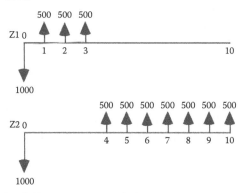

4.7 Which of the following two cash flow scenarios (Z1 and Z2) is better from the point of view of the measures of PW, AW, IRR, PBP (discounted and nondiscounted) and BCR? Use a 10% per annum interest rate for the calculations. Comment on why the comparison turns out the way it does.

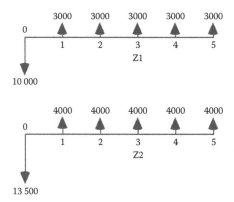

Now consider the incremental cash flow (Z2 – Z1). What do incremental PW, incremental IRR and incremental BCR measures tell you as to which is the better investment?

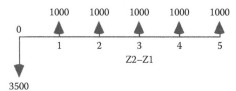

What do you conclude about the information benefit:cost ratio, annual worth, present worth, internal rate of return and payback period measures give you in establishing investment preference?

4.8 How much of the confusion between what is a benefit and what is a cost comes from the fact that for every investment there will be multiple stakeholders? Each stakeholder looks at an investment from a different perspective, and therefore may think about costs and benefits differently.

4.9 The Pareto or 80-20 rule would say that 80% of the items would contribute 20% of the amount. How could this rule be used to facilitate appraisals?

4.10 What time horizons (lifespans) do you think should be used in governmental and quasi-governmental studies? Specifically, consider the government as acting in the best interests of its citizens, both now and into the future.

4.11 How accurate are your estimates of interest rates, costs, benefits and lifespans when predicting 1, 5, 20 and 50 years ahead?

Hence, how good is a recommendation for investing in something based on assumed interest rates, costs, benefits and lifespans?

4.12 Garbage-in, garbage-out (GIGO) is a common expression in engineering analysis. It refers to the quality of the data used. The output of any analysis is only as accurate as (or less accurate than) the analysis input. Given GIGO, comment on the accuracy of any appraisal considering:

a. The magnitude of uncertainties implicit in cash flow estimates, and the timing of these cash flows.
b. The interest rates that will apply over the life of the investment.
c. The effect of taxation.

4.13 In dealing with the uncertainty in estimates (forecasts) for benefits, costs, interest rates and lifespans, could you think the same as with share trading, namely take the view that in the short/medium term the price may deviate from that forecast, but in the long term the effect of short-term fluctuations is immaterial?

4.14 In looking at changes in benefits, costs, interest rates and lifespans over the duration of an investment, would it help to distinguish the nature of the change, for example between gross changes and minor fluctuations, and also between long-term irreversible changes and short-term fluctuations? For example if it is a technological change that sharply reduces the cost, then the effect is long term.

4.15 The draft environmental impact statement on an airport extension project claimed that the extension would generate many thousands of jobs. Apart from the construction jobs, no additional jobs were known to have been created as a direct result of the airport extension. Indeed, jobs were shed in recent times. Comment on this impact statement in relation to the use of estimates (forecasts) to manipulate appraisals of investments.

4.16 What practices other than those mentioned in Section 4.4 could you adopt to deal with uncertainty in benefit–cost calculations?

4.17 Currently, how is the interest rate chosen for the following:
- The public sector, including your employer if relevant?
- The private sector, including your employer if relevant?

4.18 What is the weighted average cost of capital (WACC)? How is it defined? Is it used, or is it only something that people talk about?

4.19 What is the social opportunity cost (SOC) of capital? How is it defined? Who uses it?

What is the social rate of time preference (SRTP)? How does it relate to a social discount rate?

4.20 How far into the future should you attempt to carry the evaluation in terms of sustainability? Should each generation in the future be given equal weight in the analysis?

4.21 Within an appraisal, how should you account for technological development in the future?

4.22 The choice of interest rate reflects the weight you give the future compared to the present. Is this a philosophical issue? How does this relate to the notion that, given the choice, people prefer instant gratification rather than waiting for some future benefit?

4.23 Consider a country's nonrenewable resources and their usage or depletion. Which promotes the most usage, a low or high interest rate?

4.24 How is the notion of sustainability linked to interest rates? (Generally, any discussion of the impact of anything in the future typically refers to the future benefits. High and low interest rates value future benefits differently.)

A low interest rate gives weight to future events. From an environmentalist's viewpoint, is a zero rate equivalent to not touching the environment, because everything in the environment has some small value?

From a mining viewpoint, in terms of nonrenewable resources, is a zero rate equivalent to not doing any mining, because the largest benefit is obtained by exploiting the resource in the future and not now?

4.25 The use of high interest rates could lead to an 'intergenerational bias' when conducting appraisals. This would particularly be the case for large infrastructure projects (such as tunnels and public transport), which have high immediate costs and provide ongoing benefits over very long periods of time. Do future generations miss out on the benefits of such projects because appraisals made today with large interest rates don't demonstrate viability?

If the benefits of projects are required to be discounted at a rate of 10% per annum or more, then only those with quick payoffs will be considered, and long sustained benefits will count for little.

In view of the fact that shortsightedness by planners is universally condemned, it appears that the use of a high interest rate policy is inconsistent with community expectations. Why not use an alternative interest rate – a social time preference rate (say, a few per cent per annum) – which embodies moral judgements about the welfare of different generations? Why not use such a rate in the evaluation of long-term infrastructure projects?

4.26 Consider the effect of having a zero interest rate. Would this make more investments viable, resulting in more investment, meaning more usage of natural resources? Would depletion of natural resources follow? Or not? More investment, using more resources to produce more goods to meet the demand in a market will eventually reach a point where supply matches demand. Any more investments coming online will tip the balance toward oversupply, lower prices and a reduction in the number of viable investments. If this point is within the renewable capacity of the resource then could you consider the market to be operating sustainably?

By using a zero interest rate, would future generations be protected at the expense of the current generation? Would this, in effect, address current world issues, such as poverty, that exist around the world?

4.27 Should notions of sustainability separate economy (man-made 'capital') from environment (natural 'capital') and as such apply different interest rates to each? For example, a zero rate applied to natural 'capital' to ensure preservation for future generations, and a normal rate applied to man-made 'capital' to stimulate investment and business activity.

Or can natural 'capital' and man-made 'capital' be considered as a whole (total 'capital'), with gains in man-made stock substituting for loss of natural resources, such that total 'capital' remains constant? Would a rate higher than zero see losses in natural resources balanced by growth in man-made 'capital'?

4.28 What does using a negative interest rate mean? Does it imply that the quality of the environment will deteriorate over time? Does such a rate preserve the environment? Where do the notions of consumption and sustainability fit here?

4.29 What does a zero interest rate mean? Would you use a zero rate, for example, when the environment needs to be modified and then rehabilitated afterward, such as on mine sites where the vegetation is cleared, and then replaced after use? Does this mean that any damage done by the present generation needs to be offset by future repair?

4.30 If society is committed to sustainability, why doesn't society insist that government departments use 0% per annum or low rates for appraisal studies? Or is this where politics intrudes on rational thinking?

The appropriate rate to use should reflect society's preference for allocating natural resource usage over time. Why might the determination of a social discount rate be controversial?

4.31 The choice of interest rate is always a matter of debate. For private sector work, should it be

- The minimum acceptable rate of return for the investment?
- The opportunity cost sacrificed by not investing in an alternative?
- The weighted cost of capital used to finance the investment?
- A combination of the above?
- The above plus other things?
- Other?

What about incorporating something for risk within the rate?

4.32 How does the financial status of an investor affect the choice of interest rate?

4.33 How do you appraise an investment when financial, technical, legal, environmental, social, sustainability and political objectives/criteria are all present?

4.34 Why do transport projects get justified/not justified using predominantly revenue from ticket collections or tolls? Surely, greater utilization of a public transport system and having less reliance on roads would make more sense.

There are a number of road projects, being undertaken at the present time, that are high cost projects. In part, the projects rely on revenue streams from tolls to provide funding. Should the appraisals include comparing an equal investment in public transport?

4.35 Assume that an investment has community involvement at the appraisal stage. How do you rank or weigh the objectives/criteria of all the stakeholders in such a case?

A developer may be solely interested in profitability, the government may be interested in influencing developments, environmental protection and voters, while the public may be interested solely in protecting local concerns. How do you appraise an investment in such a case?

Can voting overcome any dilemmas? Discuss.

What role does mediation, negotiation and bargaining have in resolving such dilemmas?

4.36 How do you estimate the monetary value of the following? Is there any rationality in any approach adopted?

- The resumption of a person's house (perhaps in order that a road can be widened)? (Refer to the movie *The Castle*, Village Roadshow, 1997, where a distinction was made between a house and a home.)
- Dislocation of a community by routing a road through the middle of a neighbourhood?

- The loss of habitat of an endangered species of flora or fauna?
- A national park? A suburban playground? A tree?

4.37 In land resumption, what do you do in the case where only part of the land is subject to compulsory acquisition? For example, the widening of a road may only require part of the land to be resumed and the owners can still remain in their house, but with a reduced land area. Is it then fair to only pay market value for the proportion of the land which is resumed? Or should the owners be entitled to more compensation (proportionally) than if the whole house was subject to resumption?

An associated scenario is easement creation. The land isn't bought, but rather the right to install utilities in the land is acquired. This reduces the land use to the owner and hence its value, and so the owner must be compensated somehow.

4.38 Accountants and financiers might handle investment externalities and intangibles by ignoring them totally. On the other hand, opponents and proponents of an investment may value externalities very highly. For example on a rail link proposed to replace motor vehicle movements, the outcomes would relate to reductions or increases in air pollution, greenhouse gas emissions, noise pollution, accidents and road damage. How do you reconcile the views of the two disparate groups of people particularly when, for example, changing the valuation of external benefits can change the benefit:cost ratio and favour/disfavour different alternatives?

4.39 There is a view that intangibles cannot be given a monetary value. This is held by people who value nature and humankind, and who may even be spiritual (without necessarily being religious). The accountant's and economist's view might be that everything can be given a monetary value. Where do you lie?

4.40 There have been suggestions that money should not be used as the unit of measurement in appraisal. For example, in agriculture why aren't crops measured in terms of the quantity of water consumed in their production, and value the different crops accordingly? Comment on the use of a nonmonetary measure such as water.

4.41 If things are regarded as 'priceless' or 'invaluable', will they end up being regarded as worthless (that is, worth zero dollars) if they are not at least given some value?

4.42 What influence do carbon credits and carbon trading have on the way environmental intangibles are valued?

4.43 Is it likely that a carbon-trading scheme will have a flow-on effect to other environmental resources such that organizations begin to examine their water and material use together with waste production, and then place values against these? It would be nice to think that the high profile of carbon trading will also bring environmental intangibles into the spotlight with a positive effect.

4.44 Would trading in pollution credits be the best way to curtail emissions of all pollutants? Will emissions other than carbon be treated similarly in the future if or when they start to threaten the future of the planet?

Legislation already exists to protect the environment from many pollutants. Why can't carbon emissions be controlled in the same way? It seems that legislation is used to control your everyday backyard type of pollution, but when a pollutant starts to threaten the planet, a market to trade in is developed.

4.45 How does triple bottom line (TBL) and similar reporting treat intangibles?

4.46 How is triple bottom line and similar reporting dealt with in terms of
   a. Accounting standards?
   b. Reporting for stock exchange purposes?

There is a trend toward greater transparency and accountability in public reporting and communication, reflected in a progression toward more comprehensive disclosure of corporate performance to include the environment, social and financial dimensions of companies' activities. This trend is being largely driven by stakeholders, who are increasingly demanding information on the approach and performance of companies in managing the environment and social/community impacts of their activities in order to obtain a broader perspective of their economic impact. How this is reported to stakeholders or a stock exchange raises a series of complicated questions.

4.47 How do you take geographical/distributional effects into account in an appraisal?

4.48 Ideally, the processes of societal or public decision making should try to maximize the interests of the community at large rather than just those of the investment's proponent.

Consider the case of the construction of an extra runway at a city airport, as a way of easing the burden on the existing runways. Such an undertaking could anticipate intense public and local government opposition.

Which significant societal groups would be impacted positively by such an undertaking, and which significant societal groups would be impacted negatively? For each group, indicate the type of impact and the degree to which you feel the magnitude of the impact could be measured in financial terms.

Part II

Probabilistic

Do not go where the path may lead; go instead where there is no path and leave a trail.

Ralph Waldo Emerson

Part II

Probabilistic

Chapter 5

Background

5.1 INTRODUCTION

Uncertainty is an inherent characteristic of investments. Causes of this uncertainty lie in the cash flows, interest rates and investment lifespans. And the only rational way to incorporate this uncertainty into an investment appraisal is through a probabilistic analysis. The analysis is still referred to as *discounted cash flow (DCF) analysis*, but now is probabilistic. This chapter outlines the background material necessary for the probabilistic (embodies uncertainty) case of investment appraisal. Subsequent chapters rely on this material.

Directly incorporating uncertainty, rather than assuming determinism (even with adjustments and modifications), provides a more comprehensive investment analysis, particularly for long-term investments. Understanding the implications of the uncertainty in an investment is paramount in investment decision making. Part II of this book addresses this, by giving a readily understandable approach that provides insight on uncertainty associated with an investment.

Conventional investment appraisal for infrastructure and assets tends to be carried out deterministically, commonly using deterministic present worth (PW), where the future is assumed to be a continuation of the present, or trend from the present, such that all costs, benefits, interest rate and investment lifespan are assumed known (that is, certainty is assumed). Annual worth (AW), future worth (FW), internal rate of return (IRR), payback period (PBP) and benefit:cost ratio (BCR) are also used, sometimes in combination. Uncertainty is ignored, in spite of common acknowledgement of the existence of the presence of uncertainty. New investments are undertaken only if the present worth of the positive cash flows outweighs the present worth of the negative cash flows. If the deterministic present worth of all cash flows is positive, then the investment is viable, and if the deterministic present worth is negative then the investment is not viable.

A deterministic analysis might be supplemented with sensitivity, 'what if', or scenario analysis in an attempt to incorporate uncertainty or variability.

However such approaches are unable to differentiate grades of separation from any benchmark such as a mean value. They do not realistically consider the variability in and utilize all available information about the analysis input variables. Sensitivity analysis tries to quantify the elasticity in measures, such as present worth, by considering finite changes in the most influential analysis input variables, but does not recognize the likelihoods associated with the changes or characterize the real uncertainty in the analysis input variables. The probabilistic approach fully utilizes the available information about an uncertain investment and facilitates its correct transformation into an evaluation of an investment's true financial merit. This probabilistic form establishes a relationship between the values taken by a measure (such as present worth) and the probabilities of these values, whether the investor is risk averse, risk prone or risk neutral.

The true probabilistic nature of the uncertainty or variability should be acknowledged in order to understand the investment appraisal better. In the probabilistic case, investments may turn out as losses and also turn out as gains, with probabilities attached. In the deterministic case, the situation is more black and white – the investment is either viable or not viable, for example the present worth is either positive or nonpositive.

Where one or more of the analysis input variables of

- Cash flows
- Interest rate
- Investment lifespan

contain uncertainty and is assumed to be probabilistic, this leads to the measures of present worth, annual worth, future worth, internal rate of return, payback period and benefit:cost ratio being random variables (see Figure 4.1). The confidence with which investors can place on the outcome of any DCF analysis depends on an acknowledgment of the uncertainty which exists in these assumed analysis input variables.

Deterministic discounted cash flow analysis (Part I of this book) is well established as a tool for evaluating an investment's viability. However, writers and practitioners have long acknowledged the presence of uncertainty associated with the main analysis input variables, and the need for tools to establish and understand the risk associated with investments. A probabilistic approach permits this. Carmichael and Balatbat (2008a) provide a comprehensive survey of writing on probabilistic approaches.

The term risk is used in the sense of Carmichael (2004, 2013) to mean the exposure to the chance of occurrences of events adversely or favourably affecting the investment as a consequence of uncertainty. And not in any other sense. Risk only exists in the presence of uncertainty.

For different analysis purposes, the variables of cash flows, interest rate and investment lifespan might be allowed to be probabilistic one at a time, or together, with the other variable(s) remaining deterministic, or all together. Clearly the analysis gets more involved with less determinism in the formulation. The probabilistic version also raises the issue of uncertainty over the timing of the occurrence of cash flows.

Much of the discussion below is in terms of present worth, as the dominant form of appraisal measure. Extensions to the internal rate of return, payback period and benefit:cost ratio thinking follow.

The chapter is structured into giving background on the book's preferred method of probabilistic analysis (Section 5.2), how the appraisal measures change for the probabilistic case (Section 5.3), where options fit within probabilistic analysis (Section 5.4) and how raw data feed into the analysis (Section 5.5).

5.2 PREFERRED ANALYSIS FRAMEWORK

5.2.1 Outline

Within probabilistic discounted cash flow analysis, the analysis input (random) variables requiring probabilistic characterisation are cash flows, interest rate and investment lifespan. From these analysis input variables, the probabilistic discounted cash flow analysis gives the probabilistic characterization of the measures of PW, IRR, PBP and BCR, which themselves are also random variables. Figure 4.1 shows the transformation.

There are three main possibilities for undertaking this analysis:

- A closed form analysis using probability distributions to describe the analysis input random variables
- Monte Carlo simulation
- A second order moment analysis

The first possibility is regarded as being mathematically intractable on top of there generally not being available knowledge of the distributions of the analysis input random variables. The second possibility similarly requires knowledge of these distributions. It is also numerical and hence general conclusions cannot be drawn; while able to perform any required analysis and being powerful in this respect, it gives no insight into the analysis or understanding of investments. This book adopts the third possibility – a second order moment analysis, which is described in the next section.

5.2.2 Second order moment analysis

To perform a probabilistic analysis, the assumption is made that probabilistic information is available on the analysis input variables, though obtaining good data represents a considerable obstacle. Data leading to knowledge of the complete probability distributions of the analysis input variables are usually lacking. This constrains the closed form analysis and Monte Carlo simulation, mentioned in the previous section, to working with guesses for the probability distributions for the analysis input variables. Having said that, many people are comfortable working with Monte Carlo simulation, and so mention is made of this method in later chapters where it might be used.

A more realistic and simpler approach, and the one adopted in this book, is to use a second order moment analysis, only working with expected values and variances to characterize the random variables rather than complete probability distributions. (Deterministic variables are characterized by their expected values only and have zero variances.) This permits a ready understanding and insight into all possible investment configurations. It is easy to understand, and the associated computations are straightforward. It provides a unifying framework. Such an analysis involves the usual mean or expected values (that are used in the deterministic version of present worth), as well as variances (to incorporate uncertainty information). It is a straightforward extension of deterministic discounted cash flow analysis. The mathematical background requirement is very mild and should be understandable to those familiar with conventional deterministic analysis. It does not require any assumptions to be made on the distributions of the analysis input variables, only estimates of their moments (expected values and variances, and sometimes covariances or correlations). These are then used to obtain moments of the random variables of present worth, internal rate of return, payback period and so on.

Note, however, that in the final line of any calculations, it may be necessary to assume a distribution of the present worth, internal rate of return or payback period. With these also described in terms of their expected values and variances, as a result of the analysis, any 'two-parameter' distribution, such as the normal distribution, may be fitted.

In conjunction with only using expected values and variances to characterize each random variable, all investments are interpreted in terms of a collection of cash flows; the particular cash flows vary from case to case, but the general formulation remains unchanged. Because uncertainty is embodied directly, the notion of a discount rate adjusted for risk disappears, and it becomes appropriate instead to use a rate unencumbered by notions of risk. Expected values and variances of the input variables translate into an expected value and variance of the present worth. Information on the present worth establishes the investment viability, or in the case of options, the option value. No more knowledge than this is needed to do all investment appraisals.

The appendix to this chapter gives most of the required fundamental results for general second order moment analysis.

5.2.3 Discrete time

The analysis adopted in this book discretizes the time interval, rather than assuming continuous time. This is in line with conventional investment analysis, which is done based on days, weeks, months, etc. (periods). A spreadsheet is all that is needed to perform the calculations. Generally, the assumption is made that interest is compounded once per period, or amounts are discounted once per period (discrete time discounting), though this can be readily modified.

Equation (5.1a) gives the compound interest expression,

$$S_n = P(1 + r)^n \qquad\qquad (5.1a)$$

which assumes that interest is compounded once per period (commonly, a year). Here, P is the present value, and S_n is the equivalent future amount of P accruing at a rate r for n periods.

For compounding c times per period, then,

$$S_n = P\left(1 + \frac{r}{c}\right)^{cn} \qquad\qquad (5.1b)$$

When interest is compounded continually (continuous compounding), that is c→∞, the compound interest expression becomes,

$$S_n = Pe^{rn} \qquad\qquad (5.1c)$$

(For continuous time discounting, P is expressed in terms of S_n.)

Example

Table 5.1 shows the influence of different compounding assumptions.

From Table 5.1 it can be seen that the assumption relating to compounding (discrete versus continuous time) has some effect, but not large.

Table 5.1 Future amount with different compounding assumptions

c	S_n ($) (r = 5% p.a.)	S_n ($) (r = 10% p.a.)
1 (year)	1.05	1.10
2 (six monthly)	1.050625	1.1025
4 (quarterly)	1.050945337	1.103812891
12 (monthly)	1.051161898	1.104713067
52 (weekly)	1.051245842	1.105064793
365 (daily)	1.051267496	1.105155782
Continuous	1.051271096	1.105170918

Note: P = $1; n = 1 year.

5.2.4 Notation

The main notation adopted for the probabilistic Chapters 5 to 11 is as follows:

i	time or period counter, i = 0, 1,..., n; time may be measured in any unit, for example a day, a month or a year
n	lifespan
r	interest rate (expressed as a decimal, for example a rate of 5% per period is expressed as 0.05)
X_i	net cash flow at time i, i = 0, 1, 2,..., n
Y_{ik}	cash flow component k, k = 1, 2,..., m, in period i, i = 0, 1, 2,..., n
P[]	probability of the contained argument
E[]	expected value, mean
Var[]	variance (standard deviation squared)
Cov[]	covariance
ρ	correlation coefficient
f()	probability density function
PW	present worth
IRR	internal rate of return
PBP	payback period
BCR	benefit:cost ratio
Φ	feasibility
CDF	cumulative distribution function
PDF	probability density function

5.3 MEASURES OF VIABILITY AND PREFERENCE

5.3.1 Outline

With the analysis input variables of cash flows, interest rate and investment lifespan being random variables, it then follows that the measures of PW, IRR, PBP and so on are random variables. The second order moment analysis results in these measures being characterized in terms of their expected values and variances.

It is at this last point in the analysis that some assumption may be necessary on the shape of the distribution describing each measure, in order to obtain numbers for decision-making purposes. What is the best characterization for each measure has fortunately been studied in the literature (Carmichael and Balatbat, 2008a). This section describes the consensus view on the best distributions to use for characterizing the different appraisal measures. However the decision maker can use other distributions at his or her discretion, because the analysis framework is independent of any assumed distribution shape.

It is shown that by introducing a term *feasibility*, related to probability, all measures can be unified.

5.3.2 Probability distribution for present worth

For present worth resulting from a series of cash flows, a normal distribution appears reasonable (Hillier, 1963, 1969; Tung, 1992; Wagle, 1967), but any distribution considered a suitable model for present worth can be used. Hillier (1963) and Wagle (1967) observe that present worth is the sum of weighted terms, where the weights are the present worth factors, and early cash flows may dominate in determining the distribution shape for present worth. Wagle however notes that the net cash flows at each period may themselves be the sum of a number of variates, and hence these net cash flows could be anticipated to approach being normally distributed. Tung (1992) performs a numerical experiment in attempting to identify the appropriateness of various commonly used probability distributions in describing the probabilistic behaviour of the present worth. Tung concludes that the adoption of a normal distribution for present worth should be acceptable. The Central Limit Theorem supports the assumption of a normal distribution when the number of additive cash flows is large, irrespective of the shapes of the distributions of the investment cash flows (though it is noted that the second order moment analysis used here only requires expected values and variances for the cash flows, and makes no assumptions on their distributions).

The shape of a normal distribution is completely defined on knowing its expected value and variance, and associated probabilities are readily evaluated using standard normal probability tables.

The equation for the probability density function of a normal distribution for a random variable X is,

$$f_X(x) = \frac{1}{\sqrt{2\pi}\sigma} \exp\left[-\frac{1}{2}\left(\frac{x-\mu}{\sigma}\right)^2 \right] \qquad -\infty < x < \infty \qquad (5.2a)$$

with parameters μ and σ. These are related to the expected value and variance as follows,

$$E[X] = \mu$$

$$Var[X] = \sigma^2 \qquad (5.2b)$$

which may be used to find the particular shape of the normal distribution in any circumstance (that is, using the so-called 'method of moments').

5.3.3 Annual worth and future worth

Annual worth is only different to the present worth by a constant (the series present worth factor, or capital recovery factor) for a constant interest rate. Accordingly, the probability distribution for annual worth is different to that for present worth by a constant. The same comment will apply to future worth. Hence, all present worth comments are relevant to any annual worth or future worth analysis.

5.3.4 Internal rate of return

Internal rate of return is related to present worth. Hence, all present worth comments are relevant to any internal rate of return analysis.

The IRR may be defined as that value of r such that PW = 0. IRR = r only if PW = 0. Thus, the probability that IRR is less than an assumed r is the same as the probability that PW is negative.

$$P[IRR < r] = P[PW < 0 \mid r] \tag{5.3a}$$

For practical computation, this approach would appear satisfactory: "it is good enough for most practical purposes" (Hodges and Moore, 1968, p. 359). Some special circumstances are noted by Hillier (1965).

The distribution for IRR may be found numerically. For each of a series of values of r, E[PW] and Var[PW] are obtained leading to a probability distribution for PW; from each distribution, a value for the cumulative distribution for IRR is obtained according to Equation (5.3a), and subsequently the probability density function for IRR is obtained either by differentiation of the cumulative distribution function, or by assuming IRR follows a normal distribution. It is argued by Hillier that if the probability distribution for PW is normal, then so the probability distribution for IRR will approximate that of a normal distribution.

5.3.5 Payback period

Payback period may be similarly obtained as for internal rate of return. For each lifespan in a range of lifespans, E[PW] and Var[PW] are calculated, and a normal distribution is fitted to these. Since

$$P[PBP > \text{nominated } t] = P[PW < 0 \mid \text{nominated } t] \tag{5.3b}$$

then the cumulative distribution function for PBP is obtained from $1 - P[PBP > t]$.

5.3.6 Benefit:cost ratio

Tung (1992) performs a numerical experiment in attempting to identify the appropriateness of various commonly used probability distributions in describing the random behaviour of the benefit:cost ratio. Tung concludes that the adoption of a lognormal distribution for the benefit:cost ratio should be acceptable, where the benefits are positive.

The equation for the probability density function of a lognormal distribution for a random variable X is,

$$f_X(x) = \frac{1}{\sqrt{2\pi}\zeta x} \exp\left[-\frac{1}{2}\left(\frac{\ln x - \lambda}{\zeta} \right)^2 \right] \qquad x \geq 0 \qquad (5.4a)$$

with parameters λ and ζ. These are related to the mean and variance as follows,

$$E[X] = \exp\left(\lambda + \frac{1}{2}\zeta^2 \right)$$

$$Var[X] = \exp\left(2\lambda + 2\zeta^2 \right) - \exp\left(2\lambda + \zeta^2 \right) \qquad (5.4b)$$

or

$$\lambda = \ln E[X] - \frac{1}{2}\zeta^2$$

$$\zeta^2 = \ln\left(1 + \frac{Var[X]}{E^2[X]} \right) \qquad (5.4c)$$

which may be used to find the particular shape of the lognormal distribution in any circumstance (that is, using the so-called 'method of moments').

Other lognormal distribution uses. The lognormal distribution is applicable where negative values of a variable are not allowed or cannot occur.

If there is the possibility of present worth going negative (and this may occur wherever a negative cash flow is involved), then a lognormal distribution for present worth may not be appropriate. To apply a lognormal distribution to present worth (where negative values are possible) would require shifting the origin in order to allow negative present worth.

5.3.7 Feasibility

The notion of feasibility, as a probability, provides a unifying thread to the measures of present worth, annual worth, future worth, internal rate of return, payback period and benefit:cost ratio. Feasibility refers to an investment being worthwhile. It establishes the suitability of an investment. Its definition changes slightly between the measures of present worth, annual worth, future worth, internal rate of return, payback period and benefit:cost ratio, but all can be related to each other.

In deterministic analyses, viability (or feasibility) is readily established, whether for single or multiple investments. However, with the inclusion of uncertainty, viability or feasibility is no longer a single transition value, but rather becomes a probability. And with multiple investments in various combinations, it becomes even less obvious.

Viability or feasibility of an investment for the deterministic case may be defined in a number of ways, typically: PW > 0; IRR > nominated r; PBP < nominated t; and BCR > 1. For the probabilistic case, feasibility (denoted Φ) of an investment, as a probability, is an extension of this and also may be defined in a number of ways:

Present worth. Feasibility is the probability that the present worth of all cash flows is positive. That is,

$$\text{Feasibility, } \Phi^1 = P[PW > 0] \tag{5.5a}$$

Equivalent definitions exist for periodic (usually annual) cash flows, and future worth.

Internal rate of return. Feasibility is the probability that the interest rate (internal rate of return, IRR) corresponding to $P[PW] = 0$ exceeds a nominated value.

$$\text{Feasibility, } \Phi^2 = P[IRR > \text{nominated } r] \tag{5.5b}$$

Payback period. Feasibility is the probability that the duration corresponding to $P[PW] = 0$ is less than a nominated value.

$$\text{Feasibility, } \Phi^3 = P[PBP < \text{nominated } t] \tag{5.5c}$$

Benefit:cost ratio. Feasibility is the probability that the ratio of the worth (present, annual or future) of the cash inflows to the worth of the cash outflows (where there are no disbenefits) exceeds 1. That is,

$$\text{Feasibility, } \Phi^4 = P[BCR > 1] \tag{5.5d}$$

These feasibility expressions can be related. Φ^1 and Φ^4 are equivalent. Internal rate of return can be interpreted as the interest rate at which the feasibility Φ^1 becomes acceptable. Payback period can be interpreted as the time at which the feasibility Φ^1 becomes acceptable. That is, the feasibility measures Φ^2 and Φ^3 are related to Φ^1, and internal rate of return and discounted payback period information can be obtained directly off present worth information.

In subsequent present worth developments, the symbol Φ is used instead of Φ^1. Φ is a measure that establishes the suitability of an investment:

- Where *competing investment choices* exist, that with the largest feasibility might be preferred. Care needs to be adopted, as in the deterministic case, where competing investments are of different scales and over different time horizons.

- With an *individual investment,* the question arises as to a level of feasibility acceptable to the investor, that is, what is an acceptable level of probability that the present worth will turn out to be positive? The answer to this will depend on whether the investor is risk prone, risk averse or risk neutral, and hence requires knowledge of the investor's risk attitude.

To understand risk, it is necessary to understand feasibility. However, feasibility is a probability, and some people may not feel comfortable working with this measure. All investments will have a finite probability of being worthwhile, and a finite probability of not being worthwhile. For the deterministic case, an investment is either worthwhile (viable, feasible) or not worthwhile; that is, the probabilities are either 1 or 0.

The investor uses knowledge of the feasibility, along with other factors such as return on investment and market conditions, to assist in the decision as to the most desirable investment. Where these factors give conflicting indications, this has to be resolved as is done in the deterministic case.

The higher the Φ value, the more desirable the investment. In options terminology, a small value of Φ (close to 0) might be considered equivalent to being far from or deep out of the money (not worthwhile); a large value of Φ (close to 1) to being far from or deep in the money (worthwhile); while a value of Φ of approximately 0.5 to being close to the money (on the border between being worthwhile and not worthwhile).

Φ may be readily evaluated where present worth follows an assumed distribution. This may be a normal distribution, or any distribution considered a suitable model for present worth can be used.

In the options analysis in this book, it is shown that the option value is related to Φ. In particular, the option value is given by the Carmichael equation (see below), which is the product of Φ and the mean of the PW upside (PW > 0). That is, knowing the distribution of PW (which derives from knowing its expected value and variance), the option value can be calculated.

5.3.8 Competing investments

With a single potential investment, viability is established by looking at the feasibility Φ as described earlier. Feasibility is a probability, and viability will depend on what level of probability the decision maker is prepared to accept. Feasibility here is a constraint (Carmichael, 2013), that is for viability, the feasibility has to be greater than some probability nominated by the investor. There are no concrete guidelines on what this probability should be;

its value will depend on whether the investor is risk prone, risk neutral or risk averse.

Where selection or preference is required among multiple potential investments, it becomes a selection between probability distributions (for each measure) as the objective functions (Carmichael, 2013). Objective functions and their extremizing for investment selection between competing alternatives may be defined in a number of ways for the deterministic case, namely maximum PW, AW, FW, IRR or BCR, or minimum PBP. For the probabilistic case, there is no longer such simplicity. When dealing with the optimal selection from competing investments, Hillier (1969) uses utility as the objective function. Conversion of objective functions to constraints is also a possibility. Portfolio approaches, mathematical programming and chance-constrained programming might also be tried.

5.4 OPTIONS

5.4.1 Outline

An option gives the holder of the option the right but not the obligation to do something – to make a choice at or before a specified date, with associated cost(s) and benefit(s). For example, a premium might be paid now in return for having the right to purchase an asset at some later date. The investor will exercise that right at the later date only if it is worthwhile to do so. Since the owner of the option is not obligated to exercise that right, the value of the option only takes into account the upside potential of the investment. The potential downside of exercising the right is not considered because the investor will not exercise that right as it is not worthwhile to do so.

Depending on what the underlying asset is, options might be classified as follows:

- Financial options
- Real options

Financial options have underlying assets of stocks and similar. The term *asset* here is used to refer to whatever the underlying is, even though the underlying might not be traded (for example, as stock or carbon credits might be) or capable of exchange, and hence strictly not an asset in the dictionary sense. An example of a nontraded underlying asset is a market index. The value of the underlying asset fluctuates over time. A financial option is termed a *derivative* product, because the value of the option depends on the value of the underlying asset and has no value by itself.

Real options have underlying assets whose values do not fluctuate over time and include infrastructure and tangible investments. A real option is the right but not the obligation to do something regarding, for example, a capital investment project. This something may take many forms and includes for example delaying, expanding or abandoning a project.

Most of the terminology on options comes from the financial options literature, which is quite extensive and involves mathematics unfamiliar to practitioners.

5.4.2 Financial options terminology

A financial option gives the holder of the option the right but not the obligation to buy (or sell) an asset at a specified and agreed price by or on a specified expiration date. Where the option allows the holder to exercise the option on or before the expiration date, this is referred to as an *American option*. Where the option is allowed to be exercised only on the expiration date, this is referred to as a *European option*.

Options based directly on the price or value of some underlying asset might be referred to as *vanilla options*. *Exotic options* are modifications of vanilla options, where buyer and seller customize the cash flow structure or payoff conditions in order to meet some specific purpose such as hedging, or risk management generally.

The right to buy (or sell) the underlying asset (at some time in the future) is attained by paying (now) what is known as the *premium*. The agreed price for buying the asset (call option) or selling the asset (put option) is called the *exercise value*, *exercise price* or *strike price*. The date, specified in the option contract, up to and including which the buying (selling) takes place, is known as the *expiration date* or *maturity*. The underlying asset price might also be referred to as the market price (or stock price if the underlying asset is stock).

There are two main types of financial options – a *call* option and a *put* option.

The call option gives the holder of the option the right to buy the underlying asset by or at a certain date for a certain price. A call option is exercised by the buyer only if the underlying asset price is more than the exercise price. The buyer then purchases the asset at the exercise price. (If the intent is to make a profit, then the underlying asset is sold in the market at the asset or market price. The difference between the buying and selling prices is the gross profit gained by the buyer. The net profit is the gross profit minus the premium paid for the option.) However, if the asset price is below the exercise price, then the buyer will not exercise the option and will make a loss equal to the premium. The buyer's downside is capped at the value of the premium, while the buyer's upside is not capped and can increase with asset price.

A put option on the other hand provides the holder with the right to sell the underlying asset on or before a certain date for a certain price. If the underlying asset price is below the exercise price then the put option will be exercised by the seller. (If the intent is to make a profit, this means that the seller will be able to sell at the higher exercise price, and then buy from the market at the lower asset or market price, resulting in a gross profit. Again, the net profit is the gross profit minus the premium paid for the option.) If the asset price is above the exercise price, then the seller will not exercise the option and will make a loss equal to the premium. The seller's downside is again capped at the value of the premium, but the upside is not capped.

For each option, there are two parties. Every option is a zero-sum game. This means that if the buyer/seller makes a certain gain, then the other to the contract must experience an equivalent loss.

Combinations of options are also possible and are designed to take advantage of certain beliefs that the investor may have regarding the movement in the price of the underlying asset. They may also be used to offset the risk associated with holding an underlying asset. A common strategy is known as a 'straddle', which involves holding both a call option and put option on an underlying asset, each with the same exercise price and expiration date. The straddle is used in situations where the investor believes that the price of the underlying asset will move substantially but isn't sure in which direction. Such combinations can be analyzed by piecing together their components.

5.4.3 Variables in financial options

The variables determining the value of a financial option are shown in later chapters to be as follows:

- The value or price of the underlying asset. This changes with time.
- The exercise value or price (deterministic).
- The time to expiration. This may be fixed or variable.
- The (risk-free) interest rate.

As well, if the underlying asset is stock, dividends may be paid on stock leading up to the time of exercising.

Estimates for these analysis input variables are needed in order to value an option. Section 5.5 discusses some possible routes that the estimating might take.

Of these four variables, only the exercise price is deterministic and is an agreed value between the parties to an option contract. Accordingly, the exercise price needs to be characterized only by its expected value (variance equals zero). The remaining three variables are all random variables. In Chapter 9, they are characterized by their expected values and variances, in contradistinction to most of the literature, which characterizes them either with probability distributions or as being deterministic. The value of the underlying asset and the exercise value are later interpreted as cash flows (inflow and outflow, but dependent on whether it is a call or put option). It is shown in later chapters that as the level of uncertainty rises in the analysis input variables, the value of the option rises.

5.4.4 Real options

Real options analysis involves the investment in real assets, in contrast to financial options, which are based on the movement over time in the price or value of some underlying financial asset such as stock.

The variables determining the value of a real option are shown in later chapters to be as follows (as a comparison or analogy with financial options, they are listed in the same order as for Section 5.4.3.):

- The cash flows (both positive and negative) that follow on exercising the option, over the remaining lifespan of the investment.
- The exercise value.
- The time of exercising the option. This may be fixed or variable.
- The (risk-free) interest rate.

In contrast with financial options, all these analysis input variables may be random variables. (That is, financial options analysis is a special case of real options analysis. This is taken up in later chapters.) In later chapters, these variables are characterized by their expected values and variances, and it is also shown that as the level of uncertainty rises in these variables, the value of the option rises.

Deterministic present worth is the existing popular measure used to establish the viability of a real investment involving an option, often using a discount rate that is adjusted for risk. (See Part I of this book.) However, apart from the questionable basis of discount rates adjusted for risk, a deterministic approach ignores the uncertainty in the investment and ignores the ability of the investor to make a choice during the lifespan of the investment. A deterministic analysis fails to consider changing future circumstances that may make a future investment more desirable than it is today. A deterministic analysis calculates a lower present worth for the overall investment when compared with real options analysis. Where options exist, deterministic present worth analysis does not consider potential upside values, which may occur with low probabilities, within projects. A real options analysis may be utilized to provide a true measure of the value of an investment subject to variable future conditions. Yet industry appears hesitant to use real options analysis directly, or keeps the enhanced value provided by an options analysis in reserve as an investment safety margin.

The analysis method given in this book should overcome this hesitancy. The analysis reduces to conventional deterministic present worth analysis when there is no uncertainty present.

Real options analysis is concerned with the value embedded within investments due to uncertainty and choice. The value that the real option adds to the deterministic present worth of the investment may influence overall investment decisions; it may convert an investment being not worthwhile to one which is worthwhile, or one which is worthwhile to one more worthwhile. As such, it is commonly argued, the value of this real option should be included when examining the viability of an investment.

Real options analysis is based on the existence of uncertainty and flexibility in investment decisions. For projects with low uncertainty or flexibility, the option value is small, and the analysis is possibly not warranted.

Each of the different types of real options, outlined in the following chapters, can be seen to contain the characteristics of either a call-style or put-style option. Their analysis contains the same elements as in financial options analysis. However, it is shown that there is no need to make a distinction between different option types.

Commonly, real options are analyzed in the literature by adapting financial options methods, such as the Black–Scholes equation (abbreviated to 'Black–Scholes' here), binomial lattices and numerical simulation of the underlying asset price movement over time. However, such an approach has many critics. Volatility, which is a characteristic of the underlying asset price movement in financial options theory, is one issue, among a number, which does not translate from financial to real options. This commentary on using existing financial options results for real options is expanded in later chapters.

By analogy with financial options, the longer the duration to option exercising, and the higher the uncertainty in the variables of cash flows, interest rate and so on, the higher the value of the option. This is because both higher uncertainty (measured by *volatility* in a financial option) and longer duration until exercising cause the potential worth of the option to rise but not the potential losses, because the downside of the option is not considered. An option protects the holder against losses, no matter how large, and rewards the holder for gains, and the higher the uncertainty the more the option is worth.

5.4.5 Options commonality

The later chapters show that it is possible to present a unified approach to all options analysis, whether the option is of the financial or real type, whether it involves buying or selling or any other distinction. In particular, the option value is given by,

$$OV = E^*[PW]$$

where the superscript * implies that the investment is made only if it is worthwhile. For options involving uncertain cash flows, interest rates and so on, that are not dependent on any conditions, this becomes,

$$OV = \Phi \times \text{Mean of PW upside}$$

This is referred to as the Carmichael equation. Here Φ is $P[PW > 0]$ and the PW upside is the area of the PW distribution to the right of the origin $(PW > 0)$. OV is an estimate of the option value. This applies to both call-style and put-style options. That is, knowing the distribution of PW, the option value can be calculated. There is no need to distinguish between the different real option types such as defer, expand or abandon. There is no need to distinguish between the different types of financial options. By using the Carmichael equation, all of the different option types, whether financial or real, can be put under a single framework.

Outline derivation of the Carmichael equation. Let anything favourable to the investor be a cash inflow and anything unfavourable be a cash outflow. That is, no distinction is made between buying and selling; rather, each investment is interpreted from the viewpoint of the investor, not in any strict accounting sense.

Consider an option that can be exercised at time T. All cash outflows and inflows at and beyond T are converted to their present worth at T. The present worth at time T is,

$$PW_T = (\text{PW of all cash flows at and beyond T})_T$$

Values lying in the positive part of the distribution for the present worth correspond to worthwhile investments.

Define the option value at time T, OV_T, as the expected value of PW_T, evaluated on the assumption that the investment is only made if it is worthwhile (denoted with *). Then,

$$OV_T = E^*[(\text{PW of all cash flows at and beyond T})_T]$$

This can now be discounted to give an option value at time 0,

$$OV = pwf \times OV_T$$

$$= E^*[pwf \times (\text{PW of all cash flows at and beyond T})_T]$$

$$= E^*[PW]$$

OV is evaluated assuming that the investment is only made if it is worthwhile (in the money). For cash flows, interest rates and so on not dependent on any conditions, the Carmichael equation results.

Present worth distribution. In getting to the distribution for PW, this book adopts a second order moment approach, but Monte Carlo simulation could equally well be used if the investor prefers. The second order moment approach does not require any assumptions to be made on the distributions of the variables of cash flows, interest rates and so on. Each potential investment is reduced to its respective cash flows, and analyzed based on these. A spreadsheet is all that is needed to perform the calculations. At the end of the calculations, a distribution is fitted to the calculated present worth expected value and variance, and this distribution can be whatever the investor thinks is most appropriate, for example a normal distribution. (See Section 5.3.)

In the Carmichael equation, to calculate Φ and the mean of the present worth upside, formulae based on the equation for the distribution adopted for present worth can be used. Alternatively, for an approximate value, the upside part of the present worth distribution can be divided into vertical strips and its area and centroid calculated as a structural engineer would calculate for a member cross section. For strip s, s = 1, 2, ..., S, of width Δ, height h_s (obtained by evaluating the probability density function),

$$\Phi = \text{PW upside area} = \sum_{s=1}^{S} h_s \Delta$$

$$\text{Mean of PW upside} = \frac{\sum_{s=1}^{S} h_s \Delta}{S\Delta} \qquad (5.6)$$

This is readily evaluated on a spreadsheet. The number of strips used will be determined by whatever accuracy is desired.

Nonoption relationship. For the case involving no options, that is all cash flows in a potential investment are assumed to occur, $E^*[PW]$ becomes $E[PW]$, because there is no choice or discretion involved.

5.5 OBTAINING ESTIMATES

5.5.1 Outline

In performing a probabilistic appraisal, whether including an option or not, there is a need to characterize the analysis input variables of cash flows (or asset value and exercise value in financial options), time of exercising (if an option), interest rate and investment lifespan. In later chapters, this characterization takes the form of expected values and variances, that is in terms of moments. Should a variable be deterministic, then its variance is zero. Relationships between variables are characterized by covariances or correlations.

The following sections suggest some ways by which expected values and variances for the various variables, and their correlations, may be obtained.

5.5.2 Estimating moments

Estimating expected values and variances (moments) of any variable may be done in any reasonable way, but in the absence of anything else, the following approaches, among others, might be tried:

- First, optimistic (a), most likely (b) and pessimistic (c) values are estimated as is done in PERT. This then gives, expected value or mean = $(a + 4b + c)/6$, and variance = $[(c - a)/6]^2$. (See for example, Carmichael, 2006; Carmichael and Balatbat, 2008a.)
- As an example of this, the optimistic estimate might be taken as the value at the upper fifth percentile of the probability distribution. The pessimistic estimate might be taken as the value at the lower fifth percentile, with the most likely estimate taken as the mean of the distribution.
- A proxy approach to estimating variance can be used if the investor has previous similar investments, by analyzing the variances of similar variables.
- Investors may prefer to estimate maximum, minimum and most likely values, use a triangular distribution, and calculate an expected value and variance based on this.
- Dandy (1985) first estimates a most likely value (M), an upper value (U) and a lower value (L), where the upper and lower values represent 95% confidence limits. Then, mean = M, and variance = $[(U - L)/3.92]^2$.
- The estimates might be based on historical data, experience or subjective probability estimates and adjusted based on current news and forecasts of the future.
- A combination of the above might be used.

Uncertainty in the variables could be anticipated to increase with time; data will be less well known into the future. Such changes in uncertainty over time can be accommodated by assuming larger variances with time.

5.5.3 Estimating correlations

Data would generally not be available on the covariances or correlations of cash flows and other variables, or if it was available, accurate values could not be anticipated. Nevertheless, suggestions for obtaining estimates for correlation coefficients between cash flows at the same time period (*component correlations*), and cash flows between different periods (*intertemporal correlations*) have been advanced in the literature. See, for example, Hillier (1969), Kim and Elsaid (1988), Kim et al. (1999) and Johar et al. (2010). Cash flow correlations are discussed in Carmichael and Balatbat (2010).

Estimates of correlations may be done in any reasonable way. At best, only approximate estimates may be available. The two cases of (1) statistical independence and (2) perfect correlation bound actual correlations and provide an envelope to the actual result.

The correlation coefficient is a normalized covariance. Between two variables X and Y,

$$\rho_{X,Y} = \frac{\text{Cov}[X,Y]}{\sqrt{\text{Var}[X]}\sqrt{\text{Var}[Y]}} \qquad -1 \le \rho_{X,Y} \le 1 \qquad (5.7)$$

The correlation coefficient, ρ, is a measure of linear dependence between two variables and takes values between -1 and $+1$. Values of ρ close to $+1$ and -1 imply a linear relationship between X and Y. A small value of ρ implies a weak linear relationship, but not necessarily weak dependence. A small value does not imply that X and Y are independent. However, if X and Y are independent, ρ is zero. Perfect correlation implies ρ is $+1$ or -1. Uncorrelated random variables have a small or zero ρ value.

If values of X larger (smaller) than its mean, pair with Y values larger (smaller) than its mean, ρ will be positive. If values of X larger than its mean, pair with Y values smaller than its mean, and vice versa, ρ will be negative. That is, some dependence between X and Y exists.

This can be seen, if pairs of data exist, and the sample correlation coefficient, $r_{X,Y}$, is calculated,

$$r_{X,Y} = \frac{s_{X,Y}^2}{s_X s_Y} = \frac{1}{n} \sum_{i=1}^{n} \left(\frac{x_i - \bar{x}}{s_X} \right) \left(\frac{y_i - \bar{y}}{s_Y} \right) \qquad -1 \le r_{X,Y} \le 1 \qquad (5.8)$$

where

   n   sample size
   x_i, y_i sample values, i = 1, 2, ..., n
   $\bar{x}, \bar{y}$ sample means
   s_X, s_Y sample standard deviations
   $s_{X,Y}^2$ sample covariance

To estimate the correlation coefficient between two variables, the only way forward may be to reason logically, using physical arguments, as to what the relationship is between two variables. And then supplementing this with experience, expertise and knowledge of the situation to finally establish a reasonable estimate. For example, income this year could be anticipated to follow closely that from last year if it is based on the same asset. Hence, a high correlation between these two incomes could be anticipated. Or, the interest rate could be anticipated to generally show no relationship to a project's cash flow resulting from production in any year, and hence they could be argued to be uncorrelated. If data are available, using inbuilt correlation coefficient functions in spreadsheets might be helpful.

5.5.4 Estimating interest rate moments

Common industry practice is to assume that interest rates are deterministic, based on the assumption that interest rates do not move drastically over time. This is believed to not lose much in decision robustness. However, an examination of historical interest rates over time and the analysis presented in this book, indicate that interest rate variability does impact investment analysis, and deterministic assumptions are not giving the full story for decision-making purposes. Interest rates could be anticipated to fluctuate for a number of reasons including that due to inflation levels, exchange rates, economic reactions and central bank monetary policies.

History repeating. Central banks publish historical interest rate data, and so interest rate uncertainty over time can be readily established.

An examination of historical interest rate moments (expected value and variance) equips an investor with a strong basis for estimating the anticipated variance in the interest rate over the life of an investment; this can be then adjusted based on an investor's expertise and knowledge of the market and economy.

The time period selected to analyze interest rate data is important in establishing representative base estimates for the moments. As well, the changing economic environment needs to be considered; in recent years, these have included the global financial crisis (GFC) and the European sovereign debt crisis. This is then coupled with the investor's experience, expertise, knowledge of the market and forecasting ability, to finally establish reasonable estimates for the moments. An investment should ideally be valued using rate characteristics corresponding to the anticipated life of an investment.

Nonrepeating history. Where it is believed that future interest rates will behave differently to the past, estimating expected values and variances may be done in any reasonable way, but in the absence of anything else, the approaches suggested earlier for estimating moments might be tried.

By working only with expected values and variances of interest rates, there is the possibility of getting negative rates. (A lognormal distribution for interest rates has been suggested in the literature in order that interest rates do not go negative, even though such a distribution does not match historical data well.) However, the only situation where interest rates would go negative is if the interest rate expected value is very small and the variance is large. Zero or very small interest rate expected values might be used in sustainability or environmental arguments, but generally not commercially, and hence the likelihood of the interest rate going negative is very low.

A related comment applies to stock prices possibly going negative, if only expected values and variances are used to characterize stock prices, instead of complete probability distributions.

5.6 SUMMARY PROBABILISTIC APPRAISAL

In summary, a probabilistic appraisal goes as follows:

1. The analysis input variables of cash flows (and their timing), interest rate and investment lifespan are estimated. Because these are random variables, either their probability distribution characteristics or (as in this book) their moments will need to be established (Section 5.5).
2. Calculate the PW of the cash flows, by discounting them to present day values. This may be done via Monte Carlo simulation (which gives a histogram for PW) or (as in this book) via a second order moment analysis (Chapters 6 and 10) (which gives the moments E[PW] and Var[PW]). Fit a probability distribution to PW (Section 5.3).
3a. For nonoptions investment, feasibility Φ is P[PW > 0] (Section 5.3). Or,
3b. For an options investment, calculate the option value using the Carmichael equation (Section 5.4.5). Note that the premium is not included in the option calculation but is taken into account in the total investment viability calculation.

The procedure is almost identical for both nonoptions and options investments. The procedure differs only in Step 3, and then only in a minimal way.

It is a straightforward extension of the deterministic appraisal of Part I. Letting the variances equal zero reduces to the deterministic case. Of course, options don't exist with determinism.

Part II outline: The probabilistic present worth analysis based on second order moments is the basis of Chapters 6 to 10. Chapter 6 gives a general formulation for probabilistic cash flows. Examples illustrate the theoretical formulation. This is then adapted in order to value real options in Chapter 7. The method outlined in Chapter 7 is shown to capture the upside value of a real option in an equivalent way, and give similar results, to the Black–Scholes equation. Its strength lies in its intuitive appeal, the avoidance of having to estimate volatility, relaxed assumptions and the simplicity of the calculations. A comparison with the Black–Scholes equation is given in structural terms, with differences noted, and numerically for a range of analysis input values. Chapter 8 gives examples. Chapter 9 extends the options method given in Chapter 7 to looking at financial options. Chapter 10 incorporates probabilistic interest rates. Chapter 11 departs from the other chapters in Part II by showing how Markov chains can be used for investment analysis under uncertainty. Examples illustrate the theoretical formulations.

APPENDIX: SOME FUNDAMENTAL SECOND ORDER MOMENT RESULTS

The preferred form of analysis adopted in this book is a second order moment analysis, which underneath is an analysis of expected values and variances. Fundamental results on expectation and variance, and used in the following chapters, can be found in texts such as Benjamin and Cornell (1970) and Ang and Tang (1975). Only examples and the more important results are quoted here. All of the following (second order moment) results can be derived independently of any probability distribution assumption, thereby avoiding much complicated mathematics and avoiding having to know anything about a variable's distribution. In most cases in practice, not much more is known about a random variable than its expected value and variance. The probabilistic aspects of variables are incorporated within their variance terms.

Expectation is a linear operation, for example,

$$E[cX] = cE[X]$$

$$E[a+bX] = a+bE[X]$$

$$E[g_1(X) + g_2(X)] = E[g_1(X)] + E[g_2(X)]$$

$$E[g(X)] \neq g(E[X]) \tag{A5.1}$$

where X is a random variable; a, b and c are constants; and g, g_1 and g_2 are functions.

However, variance does not share the linear property of expectation, for example,

$$Var[c] = 0$$

$$Var[cX] = c^2 Var[X]$$

$$Var[a + bX] = b^2 Var[X] \tag{A5.2}$$

Manipulating the expression for variance and covariance, then,

$$Var[X] = E[X^2] - E^2[X]$$

$$Cov[X_1, X_2] = E[X_1 X_2] - E[X_1]E[X_2] \tag{A5.3}$$

For a linear function,

$$Z = \sum_{i=1}^{n} a_i X_i \tag{A5.4}$$

then

$$E[Z] = \sum_{i=1}^{n} a_i E[X_i]$$

$$\text{Var}[Z] = \sum_{i=1}^{n} a_i^2 \text{Var}[X_i] + 2\sum_{i=1}^{n-1}\sum_{j=i+1}^{n} a_i a_j \text{Cov}[X_i, X_j] \tag{A5.5}$$

For a product function,

$$Z = X_1 X_2 \tag{A5.6}$$

then,

$$E[Z] = \text{Cov}[X_1, X_2] + E[X_1]E[X_2] \tag{A5.7}$$

If X_1 and X_2 are independent,

$$\text{Var}[Z] = E[X_1]\text{Var}[X_2] + E[X_2]\text{Var}[X_1] + \text{Var}[X_1]\text{Var}[X_2] \tag{A5.8}$$

For a general nonlinear function,

$$Z = g(X_1, X_2, \dots, X_n) \tag{A5.9}$$

This may be expanded in a Taylor series about $E[X_1]$, $E[X_2]$, ...

Truncating this approximation such that moments no higher than variance remain, a second order approximation for expected value becomes,

$$E[Z] \approx g(E[X_1], E[X_2], \dots, E[X_n]) + \frac{1}{2}\sum_{i=1}^{n}\sum_{j=1}^{n} \left.\frac{\partial^2 g}{\partial x_i \partial x_j}\right|_{E[\,]} \text{Cov}[X_i, X_j] \tag{A5.10}$$

while a first order approximation for variance becomes,

$$\text{Var}[Z] \approx \sum_{i=1}^{n}\sum_{j=1}^{n} \left.\frac{\partial g}{\partial x_i}\right|_{E[\,]} \left.\frac{\partial g}{\partial x_j}\right|_{E[\,]} \text{Cov}[X_i, X_j] \tag{A5.11}$$

These approximations are said to be suitable provided the function is well behaved and the coefficients of variation of the X_i are not large.

Exercises

5.1 Given an expected value of –$3250 and a variance of 26 280 $², and you wish to fit a normal distribution to this, calculate:
- The values of the distribution parameters μ and σ^2
- The area under the positive part of the distribution (upside)
- The mean distance from the origin of the distribution upside (mean of upside)

5.2 Estimates for optimistic, most likely and pessimistic values of stock price are $13.86, $9.90 and $5.94 respectively. Calculate an expected value and variance of this stock price. You wish to fit a lognormal distribution to this; calculate the values of the distribution parameters λ and ζ.

5.3 Given the equation describing a normal distribution, and assuming that the associated random variable can take both positive and negative values, derive closed-form expressions for feasibility and for mean of the upside area. Check your result numerically, using numbers of your choice, with the approximation formulae given in Equations (5.6).

5.4 Considering the approximation formulae given in Equations (5.6) for feasibility and mean of distribution upside, and assuming a normal distribution, how many vertical 'strips' are necessary to get sufficient accuracy for appraisal calculations?

5.5 What distributions other than normal might be suitable to characterize present worth, where present worth can take both positive and negative values? Of these, which are asymmetric with longer tails to the right? Of the asymmetric distributions, which might be suitable for use in the analysis of stock?

5.6 Derive expressions for the expected value and variance of the function $Z = B - C$, in terms of expected values and variances of B and C. Here B stands for benefits and C costs; hence Z is similar to present worth.

5.7 Derive approximate expressions for the expected value and variance of the function $Z = B/C$, in terms of expected values and variances of B and C. Here B stands for benefits and C costs; hence Z is similar to benefit:cost ratio.

5.8 Under what conditions would Φ^1, Φ^2, Φ^3 and Φ^4 (Equations 5.5) be equal?

5.9 Would you anticipate that estimating correlations between cash flows based on your physical understanding of what the cash flows represent and their origins, to be better or worse than any mathematical approach based on numbers alone?

5.10 What distribution shape would you anticipate would best describe payback period, and why?

Chapter 6

Probabilistic cash flows

6.1 INTRODUCTION

This chapter looks at one of a number of probabilistic extensions to deterministic discounted cash flow (DCF) analysis. The analysis relaxes the deterministic assumptions on the cash flows, such that they are now random variables. That is, the cash flows are considered probabilistic, while interest rates and investment lifespans are assumed deterministic.

The general formulation given in this chapter covers many applications, with each application naturally specializing it in different ways.

Any investment is converted to a collection of cash flows characterized by their expected values and variances. The analysis then performed is a second order moment analysis. This does not require any assumptions to be made on the distributions of the investment cash flows; it only requires an assumption to be made on the distribution of the resulting present worth. A spreadsheet is all that is needed to perform the calculations.

It is remarked that Monte Carlo simulation should give similar answers. However, Monte Carlo simulation is numerical and hence provides no fundamental understanding, while it requires knowledge of the probability distributions characterizing the cash flows, and such knowledge is generally not available.

Some examples of the formulation are given.

6.2 FORMULATION

Consider a general investment, with possible cash flows extending over the life, n, of the investment. Let the net cash flow at each time period, $i = 0, 1, 2,..., n$, be the result of a number of cash flow components (random variables), $k = 1, 2,..., m$. The cash flow components can be both revenue and cost related. There may be correlation between the cash flow components at the same period.

The net cash flow X_i in any period can be expressed as

$$X_i = Y_{i1} + Y_{i2} + \ldots + Y_{im} \qquad (6.1)$$

where Y_{ik}, $i = 0, 1, 2, \ldots, n$; $k = 1, 2, \ldots, m$, is the cash flow in period i of component k, with expected value $E[Y_{ik}]$ and variance $Var[Y_{ik}]$.

The expected value and variance of X_i become

$$E[X_i] = \sum_{k=1}^{m} E[Y_{ik}] \qquad (6.2)$$

$$Var[X_i] = \sum_{k=1}^{m} Var[Y_{ik}] + 2 \sum_{k=1}^{m-1} \sum_{\ell=k+1}^{m} Cov[Y_{ik}, Y_{i\ell}] \qquad (6.3)$$

Alternatively, the variance expression can be written in terms of the component correlation coefficients, $\rho_{k\ell}$, between Y_{ik} and $Y_{i\ell}$, $k, \ell = 1, 2, \ldots, m$,

$$Var[X_i] = \sum_{k=1}^{m} Var[Y_{ik}] + 2 \sum_{k=1}^{m-1} \sum_{\ell=k+1}^{m} \rho_{kl} \sqrt{Var[Y_{ik}]} \sqrt{Var[Y_{i\ell}]} \qquad (6.4)$$

The present worth, PW, is the sum of the discounted X_i, $i = 0, 1, 2, \ldots, n$, according to

$$PW = \sum_{i=0}^{n} \left[\frac{X_i}{(1+r)^i} \right] \qquad (6.5)$$

where r is the interest rate. The expected value and variance of the present worth become

$$E[PW] = \sum_{i=0}^{n} \frac{E[X_i]}{(1+r)^i} \qquad (6.6)$$

$$Var[PW] = \sum_{i=0}^{n} \frac{Var[X_i]}{(1+r)^{2i}} + 2 \sum_{i=0}^{n-1} \sum_{j=i+1}^{n} \frac{Cov[X_i, X_j]}{(1+r)^{i+j}} \qquad (6.7)$$

Alternatively, the variance expression can be written in terms of the inter-temporal correlation coefficients between X_i and X_j, namely ρ_{ij}, rather than the covariance of X_i and X_j,

$$Var[PW] = \sum_{i=0}^{n} \frac{Var[X_i]}{(1+r)^{2i}} + 2 \sum_{i=0}^{n-1} \sum_{j=i+1}^{n} \frac{\rho_{ij} \sqrt{Var[X_i]} \sqrt{Var[X_j]}}{(1+r)^{i+j}} \qquad (6.8)$$

For independent cash flows X_i,

$$\text{Var[PW]} = \sum_{i=0}^{n} \frac{\text{Var}[X_i]}{(1+r)^{2i}} \qquad (6.9)$$

For perfect correlation of the cash flows X_i,

$$\text{Var[PW]} = \left(\sum_{i=0}^{n} \frac{\sqrt{\text{Var}[X_i]}}{(1+r)^i} \right)^2 \qquad (6.10)$$

Var[PW] is smaller for the assumption of independence compared with the assumption of perfect correlation. Perfect cash flow correlation produces a larger present worth variance and larger option value.

6.3 INTEREST RATE

An assumption is required on the interest rate. It would appear reasonable to use the cost of capital for the investor or the opportunity cost of capital or other, unadjusted for any uncertainty in the cash flows (similar to a risk-free rate), because the uncertainty in the cash flows is accounted for in the variance terms. Increasing uncertainty caused by distant time can also be accommodated in the variance estimates of the cash flows. Although the interest rate is not increased because of uncertainties in the cash flows, users might choose to increase this rate to reflect anticipated return on investment, or business practices.

6.4 PROBABILISTIC CASH FLOWS AND LIFESPAN

Where the lifespan n itself has a probability distribution, the expected values and variances referred to above, are combined over the distribution of n to give the unconditional expected value and variance of the overall present worth.

$$E[PW] = \sum_{n=N_1}^{N_2} p_n E[PW_n] \qquad (6.11)$$

$$\text{Var[PW]} = E[PW^2] - \left\{ E[PW] \right\}^2 \qquad (6.12)$$

where

$$E[PW^2] = \sum_{n=N_1}^{N_2} \left[\sum_{i=0}^{n} \left(\frac{Var[X_i] + E^2[X_i]}{(1+r)^{2i}} \right) + 2 \sum_{i=0}^{n-1} \sum_{j=i+1}^{n} \frac{E[X_i]E[X_j]}{(1+r)^{i+j}} \right] p_n \quad (6.13)$$

p_n is the probability that the investment lasts to n periods. Independence is assumed. N_1 and N_2 give the range of the distribution of n.

6.5 EXAMPLE: COMMERCIAL INVESTMENT

With the deterministic case, measures such as PW = 0 and BCR = 1 represent feasibility transition points between an investment being viable and nonviable. However, with the inclusion of uncertainty, these distinct transition points disappear, and feasibility is found to vary over the time horizon of a project investment.

Consider a project investment case example involving an industrial fabrication business. A 24-month investment horizon is examined by the investor. Optimistic, pessimistic and most likely estimates for revenue and costs are made. These are converted to expected values and variances as per Chapter 5. The variability in the expected values against time is shown in Figure 6.1. All values in the following are in $M.

The net cash flow at each time period (months), i = 0, 1, 2,..., n, is the result of two cash flow components, k = 1, 2, namely revenue and costs, $X_i = Y_{i1} + Y_{i2}$. $E[X_i]$ and $Var[X_i]$, the expected values and variances of

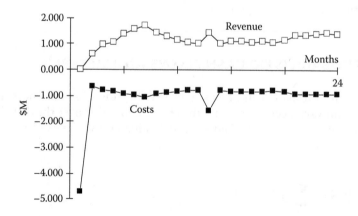

Figure 6.1 Case example – variability of revenue and costs expected values over time. (From Carmichael, D. G. and Balatbat, M. C. A., *Journal of Financial Management of Property and Construction*, 13(3), 161–175, 2008b.)

the X_i, i = 0, 1, 2, ..., n, are first calculated based on $E[Y_{ik}]$ and $Var[Y_{ik}]$. The revenue and costs are largely uncorrelated.

$E[PW]$ and $Var[PW]$, the expected value and variance of the present worth, are then calculated. The net cash flows are found to be largely uncorrelated. $E[PW]$ and $Var[PW]$ are used to construct the probability distribution (assumed here to be a normal distribution) for present worth, and the distribution for present worth is used to evaluate the feasibility Φ.

The investor expects a certain level of feasibility and also requires knowledge of the associated time, i, to reach this level of feasibility. That is, the investor would like to know the value of i that satisfies, Φ ≥ specified value.

To demonstrate the time-variant nature of feasibility and feasibility sensitivity in an uncluttered way, and so as to demonstrate the trends in the results and the type of information that feeds into the investor's decision making process, the cash flows of Figure 6.1 are approximated to an investment scenario of an initial mean outlay of $4.72M, ongoing mean cash inflows of $1.21M per month, ongoing mean cash outflows of $0.89M per month, and standard deviations set at 10% of the means. However, the calculations remain the same irrespective of any fluctuation in the means and variances of the revenue and costs. An interest rate of 0.5% per month (effective annual interest rate of approximately 6.2%) is used.

Figure 6.2 shows the change in feasibility with time.

Using deterministic calculations, the present worth would become positive at approximately 15.6 months. This is equivalent to Φ = 0.5 in Figure 6.2. The feasibility plot for the deterministic case is a step function at 15.6 months. The establishment of feasibility or viability (or not) for the deterministic case is well defined.

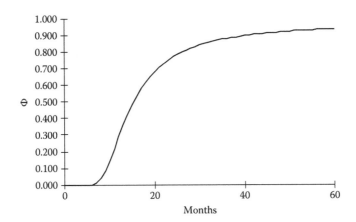

Figure 6.2 Change in feasibility with time. (From Carmichael, D. G. and Balatbat, M. C. A., *Journal of Financial Management of Property and Construction*, 13(3), 161–175, 2008b.)

However, by incorporating uncertainty in the cash flows, it is seen that the situation and the investment decision are not so straightforward. For example, if the investor is considering a 24-month time horizon, Figure 6.2 indicates that there is a finite probability (0.23) that the investor will not get a return on money invested, even though the deterministic calculations give a payback period of 15.6 months.

Assume that the investor is interested in the time by which the investment would achieve a feasibility of at least 0.75. That is, the point in time at which the probability of the cash inflows exceeding the cash outflows is 0.75. This measure, which is like a discounted payback period, can be used by the project investor, along with other factors such as return on investment and market conditions, to decide on the most desirable investment. Other measures such as feasibility related to IRR could also have been used, and where any of these measures give conflicting indications, this has to be resolved as is done in the deterministic case.

To gain a feel for the sensitivity of the results to changes in analysis inputs, Figures 6.3 and 6.4 are given.

Figure 6.3 shows the change in feasibility with change in estimates of the uncertainty in the cash flows. Different standard deviations, as a per cent of means, ranging from 5% to 25% are given. As the standard deviation increases, the time to reach a feasibility of 0.75 increases. The results indicate small sensitivity for small standard deviations, but larger sensitivity for large standard deviations.

Figure 6.4 shows little change in feasibility with change in interest rate assumptions. The results show low sensitivity to the range of interest rates exampled.

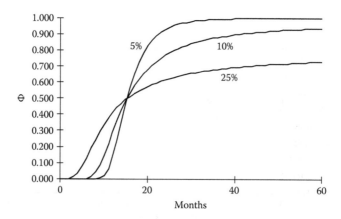

Figure 6.3 Change in feasibility with time. Sensitivity to estimates of uncertainty in the cash flows. Percentage standard deviation = 5%, 10% and 25% of mean. (From Carmichael, D. G. and Balatbat, M. C. A., *Journal of Financial Management of Property and Construction*, 13(3), 161–175, 2008b.)

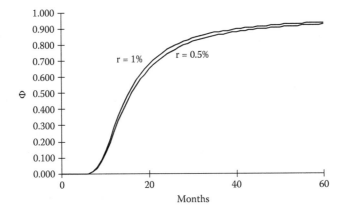

Figure 6.4 Change in feasibility with time. Sensitivity to interest rate. Monthly interest rate = 0.5%, 1%. (From Carmichael, D. G. and Balatbat, M. C. A., *Journal of Financial Management of Property and Construction*, 13(3), 161–175, 2008b.)

6.6 MANAGED INVESTMENT IN PRIMARY PRODUCTION

6.6.1 Outline

Plantations, forests, aquaculture and similar agribusiness ventures are promoted as investments contributing to sustainability, and commonly through managed investment schemes set up as businesses. With taxation concessions, investment in such ventures is popular and is perceived as being socially responsible, but it is not without risk. An analysis of business failures leads to a finite probability that the investor will lose money. The attractiveness of the investment is further diminished because of the uncertainty in the end product, which translates to uncertainty in the downstream return on investment and time of the return. Based on probabilistic arguments, a robust model and methodology on which to make investment decisions can be developed. This approach here quantifies what typically is undertaken qualitatively or deterministically.

Terminology. A managed investment scheme is an investment vehicle, and involves people (private investors or scheme members) contributing money to acquire a stake in benefits produced by the scheme. The contributions from members are pooled or used in a common enterprise. Members of the scheme do not have day-to-day influence over the operation of the scheme, but rather some nominated entity operates and manages the scheme. As such, investors rely on the effort of the nominated entity in order to get a return on their investment. The nominated entity effectively operates a business that is set up and operated using private investors' money; a company is set up for the sole purpose of growing and harvesting trees, fish, etc. Managed investment schemes cover a wide variety of investments, but here specifically addresses schemes based on primary production, such as agriculture, livestock, horticulture, forestry and aquaculture. In horticultural and forestry investment schemes, the nominated entity may be responsible for acquiring the land, planting, maintaining, harvesting and sale of the crop; the return to the investor comes on the sale of the crop and any tax concessions. In livestock schemes, the nominated entity may be responsible for acquiring the animals, looking after them and selling them; the return to the investor comes on the sale of the animals and any tax concessions. As an investment, the period of concern starts with initial investment in the scheme and concludes on the sale of the scheme's product. The terms investment and return, here, refer to money alone; social and environmental issues are not explored.

The typical cash flow for an investor involves an initial outlay, with a return at year n; alternatively the initial outlay may be spread over n years as an equivalent annual amount, and the return may occur in stages. Uncertainties in this investment scenario are in terms of when the return will be obtained, the magnitude of the return and the survivability of the scheme as a business to the point where the return is obtained. That is, n, the return, and the time to business failure are random variables. The time when the return is obtained, n, and the time to business failure in general are not the same. Investment returns are uncertain and depend on the economic conditions, extreme natural events, weather patterns, competing products, pests, diseases, consumer behaviour and so on prevailing at or up to the time of harvest of the product. The time of harvest is uncertain and will depend on management practices carried out in the preceding years, climate conditions and so on. In agribusiness investment, n can be as large as 25 years.

An analysis of businesses over time, following their initial listings, shows a decreasing probability of failure. Nevertheless, at any time in the life of a business there is always a finite probability of failure. This implies that, not only has the investor to consider return on investment, but also the likelihood that all will be lost should business failure occur. The probability of gain, and risk, from investment in agribusinesses is composed of a number of (uncertain) components – business survivability, return magnitude and return timing.

6.6.2 Business survivability

A reasonably large literature exists on business failure prediction. Failure of the product (as distinct from failure of the business) is dealt with in the probabilistic treatment of investment return later.

Consider a business subjected, over time, to causes that might bring about its demise. Such causes might be lack of capital, bad debts, insufficient return, personnel death or retirement, economic conditions and market changes amongst others. Each cause is associated with a small probability of business failure. This situation is readily dealt with using (dynamic) systems reliability theory. The probability of surviving each cause is assumed independent of surviving previous causes.

For such assumptions, the probability that the business will still exist at time t (the survivability) is

$$S(t) = e^{-\lambda t} \tag{6.14}$$

where t is time, and λ is the failure rate. That is, business survivability over time decreases exponentially with time.

In terms of the random variable T, the time to failure, $S(t) = P[T > t]$ = probability that failure hasn't occurred by t; and $S(0) = 1$; $S(\infty) = 0$.

Equation (6.14) can be shown to agree with actual business failure data, and hence could be considered a good model of failure. The parameter λ can be estimated from business failure data, and could be anticipated to be different for different business circumstances – for all businesses, for different business types, for different business failure types and for different time periods. An exact λ value is not essential because the results below show that the investor's exposure is relatively insensitive to the λ value chosen.

6.6.3 Investment return

The following development is a special case of the equations presented in Section 6.2. The scenario assumed here is that of an investor providing an initial outlay X_0, and in year n receiving a return X_n. Alternatively, the return could be spread over a few years, and the initial outlay X_0 may be spread over n years as an equivalent annual amount. Both alternatives can be accommodated in the following development. n and X_n are the random variables in this formulation. Taxation influences can be accommodated.

For the present worth, PW_n, of an amount X_n occurring in year n,

$$E[PW_n] = \frac{E[X_n]}{(1+r)^n}$$

$$Var[PW_n] = \frac{Var[X_n]}{(1+r)^{2n}} \tag{6.15}$$

where the return occurs over several years $n_1, n_1 + 1,..., n_2$, then

$$E[PW_n] = \sum_{i=n_1}^{n_2} \frac{E[X_i]}{(1+r)^i}$$

$$Var[PW_n] = \sum_{i=n_1}^{n_2} \frac{Var[X_i]}{(1+r)^{2i}} + 2\sum_{i=n_1}^{n_2-1}\sum_{j=i+1}^{n_2} \frac{Cov[X_i,X_j]}{(1+r)^{i+j}} \tag{6.16}$$

It is anticipated that X_i and X_j will be close to being perfectly correlated. In which case,

$$\text{Var}[PW_n] = \left(\sum_{i=n_1}^{n_2} \frac{\sqrt{\text{Var}[X_i]}}{(1+r)^i} \right)^2 \tag{6.17}$$

Consider the case with the single return X_n further. Where the term n itself follows a probability distribution, the expected values and variances are combined over this distribution of n to give the unconditional expected value and variance of the present worth of X_n.

$$E[PW] = \sum_{n=N_1}^{N_2} p_n E[PW_n]$$

$$\text{Var}[PW] = E[PW^2] - \left\{ E[PW] \right\}^2 \tag{6.18}$$

Here p_n is the probability that the investment goes to n years, and N_1 and N_2 are the limits of the distribution of n. Interestingly, calculations show that the investor's exposure is relatively insensitive to the distribution for n.

Figure 6.5 shows how the feasibility, Φ, varies with time, for some example values.

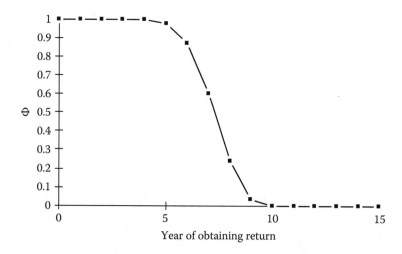

Figure 6.5 Example showing how feasibility, Φ, varies with time; standard deviations 10% of mean; $X_n = 2X_0$; $n_1 = n_2$; $N_1 = N_2$; normal probability assumptions. (From Carmichael, D. G. and Balatbat, M. C. A., International Journal of Project Organisation and Management, 3(3/4), 273–289, 2011.)

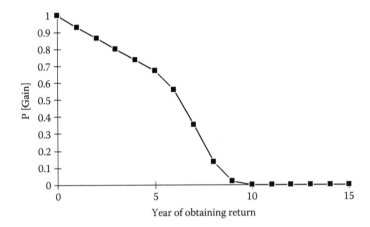

Figure 6.6 Example showing how P[Gain] varies with time; standard deviations 10% of mean; $X_n = 2X_0$; $n_1 = n_2$; $N_1 = N_2$; normal probability assumptions; $\lambda = 0.075$. (From Carmichael, D. G. and Balatbat, M. C. A., *International Journal of Project Organisation and Management*, 3(3/4), 273–289, 2011.)

6.6.4 Probability of gain

In a present worth sense, the probability of gain from an investment depends on the probability of business survival (that is, survivability) and the probability that the present worth is positive (that is, feasibility).

For probabilistic independence of survivability and feasibility,

$$P[Gain] = S\Phi$$

$$P[Loss] = 1 - P[Gain] = 1 - S\Phi \qquad (6.19)$$

Figure 6.6 shows how P[Gain] varies with time, for some example values.

6.7 CLEAN DEVELOPMENT MECHANISM ADDITIONALITY

6.7.1 Introduction

The Clean Development Mechanism (CDM) is one of the flexibility mechanisms defined in the Kyoto Protocol. For registration of a CDM project, and hence entitle the project to access saleable carbon credits known as Certified Emission Reductions (CERs), additionality must be demonstrated. Commonly, financial additionality and viability are demonstrated through deterministic internal rate of return (IRR) benchmark analysis, supplemented with a sensitivity analysis. Here financial addi-tionality and project viability are examined in the presence of cash flow uncertainty, where IRR becomes a random variable. A case study project involving wind power is exampled. It is seen that the boundaries between acceptance and rejection as a CDM project based on financial addition-ality and viability tests become blurred, leading to possible alternative conclusions.

The viability of a CDM project may be justified through a combination of the sale of the project's end-product, and the saleable carbon credits (CERs) generated by the project. Each CER is equivalent to one tonne of CO_2, with its value depending on the carbon markets. The CERs derive from emission reductions created by the project when compared with a baseline (that occurring in the absence of the project), thereby ensuring the environmental worth of the CDM project.

This section addresses the specific matter of financial additionality and viability of CDM projects in the presence of uncertainty. Other types of additionality – environmental, technology and regulatory – are not discussed.

For a project to be accepted as a CDM project, it must, among other things, satisfy a financial additionality test, which essentially involves showing that the project is not viable without the inclusion of the revenue from CERs. Common practice uses a benchmark IRR deterministic anal-ysis coupled with a sensitivity analysis. The United Nations Framework Convention on Climate Change (UNFCCC) specifies where benchmarks shall be derived from. Criticism of this practice derives from its inability to judge the degree to which a project satisfies additionality, that is, it is unable to differentiate grades of separation from the benchmark. It is also vulnerable to assumptions on cash flows; it does not properly account for uncertainty in the cash flows.

In typical additionality calculations, cash flows are assumed to be deter-ministic, and this is supplemented with a sensitivity analysis, but still the analysis remains deterministic. Such an approach is usually justified in terms of expediency. Allowing for uncertainty gives a more realistic IRR analysis and assists in identifying false positive and false negative conclu-sions on additionality.

With uncertainty in cash flows (including that from carbon credits) gen-erally, IRR becomes the more realistic characterization of being a random variable rather than a deterministic value. Accordingly, in any additionality or viability benchmark analysis, it is now a probability distribution for IRR which is involved, rather than a single number. This is shown below to imply that there is a finite probability that there will be values of IRR both above and below any given benchmark (without and with CER revenue). Expressed another way, it is possible to show that there is a finite prob-ability that all projects demonstrate financial additionality and viability, and that there is a finite probability that all projects don't demonstrate financial additionality and viability. This is an extension of the view in Chapter 5 and preceding sections in this chapter, where there is a finite probability of any project being feasible and also nonfeasible. This raises issues with additionality and viability tests.

6.7.2 Financial additionality

The means of demonstrating financial additionality is outlined in UNFCCC documents. Of the available procedures outlined, the most commonly applied appears to be benchmark analysis. This involves demonstrating that the project is not financially viable without CDM classification (and hence sale of CERs) and typically involves calculating the financial measure IRR without CER revenue and comparing against a benchmark value. UNFCCC specifies where benchmarks shall be derived from, though choice of an appropriate benchmark may have some flexibility. Projects must not be financially viable without CER revenue. In terms of an IRR benchmark (BM),

$$IRR_{withoutCERs} < IRR_{BM} \qquad (6.20a)$$

Although not part of the CDM additionality test, the situation of what happens after CER revenue is included is also of interest to investors. The introduction of carbon price estimates and CER quantity estimates adds further uncertainty to the analysis. Projects (to conventional investors) must be viable with CER revenue,

$$IRR_{withCERs} > IRR_{BM} \qquad (6.20b)$$

It is noted that CER revenue may not have a meaningful impact on some projects, that is the IRR may only slightly change on adding CER revenue, leading to questionable additionality acceptance. There also occurs erroneous rejection of projects capable of being shown to have true environmental worth, and acceptance of projects not capable of being shown to have true environmental worth. Acknowledging uncertainty directly in the analysis has the potential to address these issues.

6.7.3 Investment analysis incorporating uncertainty

Allowing the carbon price and cash flows to be random variables results in a probability distribution for IRR. Expected values and variances of cash flows, carbon price and CER quantity are estimated, leading to expected value and variance of PW, which in turn leads to a distribution for IRR.

With IRR following a normal or similar distribution, all rates have a finite probability of occurrence. Against a deterministic IRR benchmark:

- Additionality: Without CER revenue, all projects have IRRs with a finite probability of being below/above the benchmark. Both additionality and nonadditionality can be simultaneously demonstrated for all projects. All projects could be accepted, or all projects could be rejected based on an additionality test.
- Viability: With CER revenue, all projects have IRRs with a finite probability of being below/above the benchmark. Both viability and nonviability can be simultaneously demonstrated for all projects. All projects could be accepted, or all projects could be rejected based on a viability benchmark test.

6.7.4 Wind power example

Consider a wind power project. Three cash flow variance scenarios are considered in the analysis. The scenarios range from fully deterministic to fully probabilistic:

1. Deterministic estimates (zero variance), as used in existing IRR analysis
2. Uncertainty in CER price and quantity (with deterministic other cash inflow and cash outflow values)
3. Uncertainty in all cash flows

The product of price and quantity gives the cash inflow from CERs; here it is assumed that there is no correlation between price and quantity.

A deterministic IRR benchmark of 10.5% per annum is used for example purposes. Cash outflow – the pessimistic and optimistic deviations from the most likely are assumed to be 25%, being based on similar values mentioned in the literature. These deviations are tested by use of a sensitivity-style analysis. Cash inflow – the pessimistic and optimistic deviations are assumed to be the same as for cash outflow, based on published electricity prices. Balatbat et al. (2012) give an analysis of means and variances of CERs from past CDMs. Intercomponent correlation, between general cash inflows and cash outflows, is assumed to be weak but positive, based on an understanding of underlying characteristics rather than through using any mathematical formula. This assumption is tested through a sensitivity-style analysis. The correlation between general cash inflow and CER cash inflow is assumed to be strong and positive, due to their mutual connection with electricity produced, but not perfect correlation due to the lack of a link between electricity price and CER price. General cash inflow and CER cash inflow are assumed to have a small positive correlation with cash outflow, because of the relationship between electricity generation and operation costs, but not with installation and maintenance costs. Last, for this study, intertemporal correlations are assumed small and positive initially, and then tested through a sensitivity-style analysis.

With the above assumptions, the following results should be viewed as showing indicative behaviour, rather than producing specific numbers.

Scenario 1. The change in deterministic PW of the project with interest rate is shown in Figure 6.7.

At the assumed IRR benchmark rate, the project is not viable without CERs but viable with CERs. The project could be said to satisfy the CDM additionality test. And the addition of CER revenue could be considered very favourable.

Scenario 2. Refer to Figure 6.8. With the inclusion of uncertainty in the CER price and quantity only, the cumulative distribution function (CDF) plot starts to flatten, but only mildly. The without-CER case remains unchanged. Additionality could be said to hold. The without-CER and with-CER cases are markedly away from the benchmark.

Scenario 3. Refer to Figure 6.9. Considering uncertainty in all cash flows results in a marked flattening of the cumulative distribution function. Both the without-CER and with-CER cases result in values below and above the benchmark IRR. Both the without-CER and with-CER cases have finite probabilities of being both under and over the benchmark.

As the level of uncertainty in the cash flows increases in going from Scenario 1 through 3, the probabilities of being less than or greater than the benchmark change.

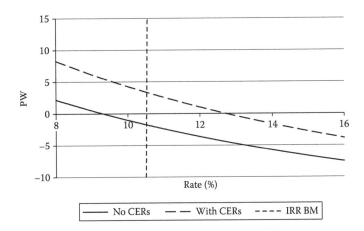

Figure 6.7 Change in deterministic PW (€×10⁶) with interest rate. (From Carmichael, D. G. et al., The Financial Additionality and Viability of CDM Projects Allowing for Uncertainty, School of Civil and Environmental Engineering, The University of New South Wales, Sydney, 2014.)

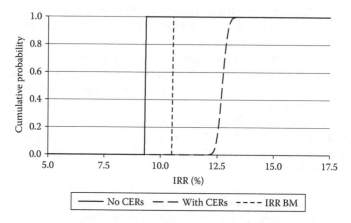

Figure 6.8 CDF of IRR under uncertainty in CER price and quantity. (From Carmichael, D. G. et al., The Financial Additionality and Viability of CDM Projects Allowing for Uncertainty, School of Civil and Environmental Engineering, The University of New South Wales, Sydney, 2014.)

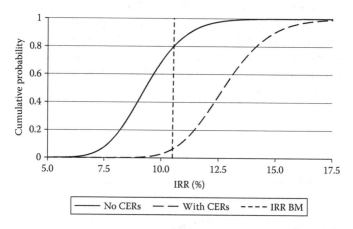

Figure 6.9 CDF of IRR under uncertainty in all cash flows. (From Carmichael, D. G. et al., The Financial Additionality and Viability of CDM Projects Allowing for Uncertainty, School of Civil and Environmental Engineering, The University of New South Wales, Sydney, 2014.)

6.8 MULTIPLE PROJECTS/VENTURES

6.8.1 Introduction

With a conventional deterministic discounted cash flow analysis, the feasibility calculations change little in going from one to many investment projects. However with uncertainty attached, the feasibility calculations need to be reworked, and the issue of feasibility becomes less transparent on going from one to many projects. The issues considered here relate to the changes in feasibility in going from one project to more than one project, and how far into the future the project cash flow should be relied upon, given that the project owner expects a reasonable level of feasibility attached to the investment.

The theoretical development given here is general. The example used to illustrate the theory makes a number of specific assumptions in order to demonstrate the ideas. It is shown that uncertainty influences the feasibility of undertaking additional projects.

6.8.2 Multiproject extension

Let the number of projects be counted by $\alpha = 1, 2,..., q$. Other notations specific to the multiproject/venture analysis is as follows:

$\overline{\Phi}$ nonfeasibility

ϕ_α the event of viability of project α, $\alpha = 1, 2,..., q$

Φ_α feasibility of project α, $\alpha = 1, 2,..., q$. $\Phi_\alpha = P[\phi_\alpha]$

$\overline{\phi}_\alpha$ the event of nonviability of project α, $\alpha = 1, 2,..., q$

$\overline{\Phi}_\alpha$ nonfeasibility of project α, $\alpha = 1, 2,..., q$. $\overline{\Phi}_\alpha = P[\overline{\phi}_\alpha]$

An investor expects a reasonable level of feasibility from a project, requires knowledge of the associated time, i, to reach this level of feasibility, and must contemplate how many projects to invest in. That is, the investor would like to know the values of i and α that satisfy, $\Phi \geq$ specified value.

Numerous assumptions can be postulated where multiple projects exist. Many of these assumptions are given below. However, it is emphasized that all assumptions may not be relevant or realistic in any given application. The investor should select the assumptions relevant to the application, and ignore the remainder.

In terms of assumptions, two broad views may be taken:

- The collection of projects is treated as a combined investment.
- The individual projects are treated as separable investments.

Within these, more specific cases can occur.

In all of the example calculations given here, the data of the project given in Section 6.5 are used. Additionally, it is assumed that the investor has specified a required 0.75 feasibility level, that is, the investor is interested in the time by which a feasibility level of 0.75 is reached. Cash flows between individual projects are assumed uncorrelated. All component projects are assumed to be the same. These assumptions are in order to demonstrate trends. The theoretical development, however, is for general assumptions.

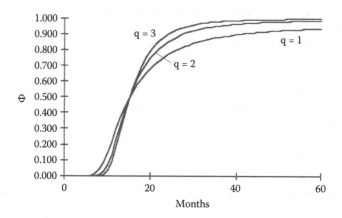

Figure 6.10 Change in feasibility with time. One, two and three simultaneous projects. Assumption – additive project cash flows. (From Carmichael, D. G. and Balatbat, M. C. A., *Journal of Financial Management of Property and Construction*, 13(3), 161–175, 2008b.)

6.8.3 The collection of projects as a combined investment

Where it is assumed that cash flows from different projects are additive, the expected value and variance of the present worth of the combined investment are obtained through combining the expected values, variances and covariances of the cash flows of the individual projects.

> **Example**
>
> Figure 6.10 gives the case where it is assumed that the project cash flows are additive. For a feasibility of 0.75, earlier payback occurs with higher project numbers.

6.8.4 The collection of projects as separable investments

Bounds or an envelope can be established on the feasibility of the collection of projects by considering two extreme cases:

- All projects in the collection individually viable.
- Only one project in the collection viable.

Each of these cases is considered in turn.

Examples are given with Figures 6.11 to 6.13 to show how the feasibility varies with time where the individual projects are separable investments.

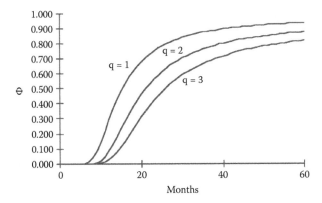

Figure 6.11 Change in feasibility with time. One, two and three simultaneous projects. Assumption – viability of all projects is necessary for total viability. (From Carmichael, D. G. and Balatbat, M. C. A., *Journal of Financial Management of Property and Construction*, 13(3), 161–175, 2008b.)

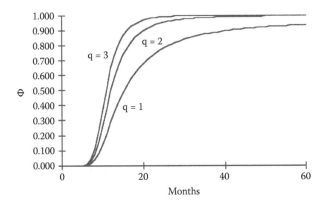

Figure 6.12 Change in feasibility with time. One, two and three simultaneous projects. Assumption – viability of one project is all that is necessary for total viability. (From Carmichael, D. G. and Balatbat, M. C. A., *Journal of Financial Management of Property and Construction*, 13(3), 161–175, 2008b.)

6.8.4.1 All projects individually viable

Where it is assumed that viability of the collection of projects depends on all projects being individually viable, and letting ϕ_α be the event of viability of project α, $\alpha = 1, 2,..., q$, then the event of viability of the collection of projects is given by

$$\phi = \phi_1 \cap \phi_2 \cap ... \cap \phi_q \tag{6.21}$$

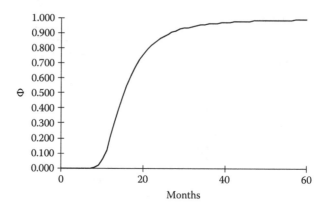

Figure 6.13 Change in feasibility with time. Three projects, with one being surplus. Assumption – viability of two projects is necessary for total viability. (From Carmichael, D. G. and Balatbat, M. C. A., *Journal of Financial Management of Property and Construction*, 13(3), 161–175, 2008b.)

and

$$P[\phi] = P[\phi_1|\phi_2 \cap ... \cap \phi_q]P[\phi_2|\phi_3 \cap ... \cap \phi_q]...P[\phi_q] \qquad (6.22)$$

For independent events, this leads to

$$\Phi = \prod_{\alpha=1}^{q} \Phi_\alpha = \prod_{\alpha=1}^{q}(1 - \bar{\Phi}_\alpha) \qquad (6.23)$$

Because each Φ_α is less than 1, this implies that the feasibility of the collection of projects is less than the feasibility of each project singly considered. Independence of events is assumed in the case example calculations below.

Example

Where it is assumed that viability of the collection of projects depends on all projects being individually viable, Figure 6.11 follows. The feasibility of the whole declines with the number of projects.

6.8.4.2 Only one project individually viable

Where it is assumed that viability of the collection of projects depends on only one project being viable, in contrast to the previous case where event viability of the collection of projects is given by the intersection of individual event project viabilities, and denoting $\bar{\phi}_\alpha$ as the event of nonviability of project α, $\alpha = 1, 2, ..., q$, then the event of nonviability of the collection of projects is given by

$$\bar{\phi} = \bar{\phi}_1 \cap \bar{\phi}_2 \cap ... \cap \bar{\phi}_q \qquad (6.24)$$

and

$$P[\bar{\phi}] = P[\bar{\phi}_1 | \bar{\phi}_2 \cap ... \cap \bar{\phi}_q] P[\bar{\phi}_2 | \bar{\phi}_3 \cap ... \cap \bar{\phi}_q] ... P[\bar{\phi}_q] \qquad (6.25)$$

For independent events, this leads to

$$\bar{\Phi} = \prod_{\alpha=1}^{q} \bar{\Phi}_\alpha \qquad (6.26)$$

or

$$\Phi = 1 - \prod_{\alpha=1}^{q} (1 - \Phi_\alpha) \qquad (6.27)$$

Feasibility of the collection of projects increases as the number of projects increases. Independence of events is assumed in the case example calculations below.

Example

Where it is assumed that viability of the collection of projects depends on only one project being viable, Figure 6.12 follows. The feasibility of the whole improves with the number of projects. But there is a decreasing relative contribution as more projects are added.

6.8.4.3 Some projects individually viable

For collections of projects comprising situations where it is assumed that the viability of several projects determines viability of the collection of projects, then the overall feasibility can be analyzed by appropriately combining the above results.

For example, consider the three-project ($q = 3$) investment case, in which it is assumed that overall viability depends on project 1 and either project 2 or project 3 being viable. Combining Equations (6.23) and (6.27) leads to

$$\Phi = \Phi_1[1 - (1 - \Phi_2)(1 - \Phi_3)] \qquad (6.28)$$

or, where the individual projects have the same feasibility, Φ_s,

$$\Phi = 2\Phi_s^2 - \Phi_s^3 \qquad (6.29)$$

6.8.4.4 Partial surplus of projects

With this arrangement, extra projects are contemplated in order to guarantee viability should some projects not achieve viability. The feasibility of such arrangements can be evaluated using the binomial distribution, based on the feasibilities and nonfeasibilities of the individual projects, Φ_α and $\bar{\Phi}_\alpha$, $\alpha = 1, 2, ..., q$.

Consider, for example, the three-project $(q = 3)$ case, where it is assumed that any two viable projects implies viability of the collection of projects. Then,

$$\Phi = \Phi_1\Phi_2\Phi_3 + \Phi_1\Phi_2\bar{\Phi}_3 + \Phi_2\Phi_3\bar{\Phi}_1 + \Phi_3\Phi_1\bar{\Phi}_2 \tag{6.30}$$

or, where the projects have the same feasibility, Φ_s, and nonfeasibility, $\bar{\Phi}_s$,

$$\Phi = \Phi_s^3 + 3\Phi_s^2\bar{\Phi}_s \tag{6.31}$$

> Example
>
> Consider having one surplus project in three, such that it is assumed that viability of the collection of projects depends on any two projects being individually viable. Figure 6.13 shows how the feasibility varies with time.

6.8.4.5 Summary: Case example

The calculations indicate for the example data that it is better to build in viability at the individual project level than at the combined project level when separable investments are present.

6.9 BENEFIT:COST RATIO

Dandy (1985) obtains approximate expressions for the expected value and variance of the benefit:cost ratio (BCR), based on expected values and variances of the period benefits and costs.

For $B = B_0 + B_1 + ... + B_n$, the expected value and variance of the total benefit are given by

$$E[B] = \sum_{i=0}^{n} B_i \tag{6.32}$$

$$Var[B] = \sum_{i=0}^{n} Var[B_i] + 2\sum_{i=0}^{n-1}\sum_{j=i+1}^{n} \rho_{ij}\sqrt{Var[B_i]}\sqrt{Var[B_j]} \tag{6.33}$$

where ρ_{ij} is the correlation coefficient between period benefits B_i and B_j.

The expected value and variance of the total cost, $C = C_0 + C_1 + \ldots + C_n$, is obtained similarly.

Approximate expressions for the expected value and variance of BCR are

$$E[BCR] = \frac{E[B]}{E[C]}\left(1 + V_C^2 - \rho_{BC} V_B V_C\right) \qquad (6.34)$$

$$Var[BCR] = \left(\frac{E[B]}{E[C]}\right)^2 \left(V_B^2 + V_C^2 - 2\rho_{BC} V_B V_C\right) \qquad (6.35)$$

where V is the coefficient of variation, and the correlation coefficient ρ_{BC} is given by

$$\rho_{BC} = \left(\sum_{i=0}^{n}\sum_{j=0}^{n} \rho_{ij}\sqrt{Var[B_i]}\sqrt{Var[C_j]}\right) \bigg/ \sqrt{Var[B]}\sqrt{Var[C]} \qquad (6.36)$$

Good correspondence with the results of others occurs particularly for lognormal or gamma assumptions on BCR.

Exercises

6.1 For the managed investment in primary production results given in Section 6.6, conduct some trends and sensitivity calculations related to magnitude of return X_n, business failure rate λ, interest rate r, degree of uncertainty (standard deviation) in the return and return time distribution. Then consider how robust the results are.

The base case is Figure 6.6 assumptions, namely – standard deviations 10% of mean; $X_n = 2X_0$; $n_1 = n_2$; $N_1 = N_2$.

a. Consider different ratios of return/initial investment of 1.5, 2 and 2.5.

b. Consider different business failure rates of 0.06, 0.075 and 0.09.

c. Consider different interest rates of 5%, 10% and 15% per annum.

d. Consider different ratios of standard deviation/mean of 5%, 10%, 20%.

e. Consider different distributions for the return – deterministic, uniform over 3 years, uniform over 5 years.

6.2 For the managed investment in primary production results given in Section 6.6, the P[Gain] curve is seen to have two transitions and three regions. Suggest a distinction between lower and higher risk investments based on these transitions.

6.3 For the managed investment in primary production results given in Section 6.6, how might the effect of any tax concessions and taxation on money earned be taken into consideration by suitably altering the values of X_0 and X_n.

6.4 For the managed investment in primary production results given in Section 6.6, suggest a summary approach for an investor to follow, in evaluating the risk associated with any managed agribusiness investment.

6.5 For the additionality calculations of Section 6.7, ideas such as an 'under-benchmark probability' (UBP) and an 'over-benchmark probability' (OBP) might be introduced. UBP and OBP have different purposes:

- UBP applies to the case without CER revenue. It is the probability that the IRR is less than the benchmark, or the probability that the additionality test is met. (For a deterministic analysis, if the project IRR is below the benchmark, the project will satisfy the additionality test, because the project is nonviable without CER revenue.) As UBP gets larger for any project, the probability of satisfying the additionality test gets larger. Amongst these projects, there may be environmentally nonworthwhile projects (false positives). The probability of a project being rejected is 1-UBP, and amongst these there may be environmentally worthwhile projects (false negatives).

- OBP applies to the case with CER revenue (which is not part of the CDM additionality test, but which is of interest to investors). It is the probability that the IRR is greater than the benchmark, that is the probability that the project is viable to an investor. (For a deterministic analysis, if the project IRR is above the benchmark, it is viable.) The value of OBP acceptable to an investor will reflect that investor's risk attitude. Risk-averse investors would like projects with high OBP values; risk-seeking or public investors would accept lower OBP values.

- Environmentalists might argue for projects demonstrating a large difference between UBP and OBP values in an attempt to maximize the reduction in greenhouse gas emissions. A large difference between the IRR expected values for the without-CER revenue and the with-CER revenue cases implies that the CERs generated by the project have a large impact.

    The issue arises as to what would constitute acceptable UBP and OBP levels. First, there is no need for the UBP and OBP requirement levels to be the same, or the same across all project types. As well, the levels can be changed depending on society's goals in terms of the number and type of CDM projects it desires.

    What do you think would be reasonable values to specify for UBP and OBP?

6.6 Based on Section 6.7. A weakness with deterministic CDM project proposals and their justification is that estimates of project output and cash flows (and emissions) can be manipulated to a certain extent, though the intent of third-party validation of CDM proposals is to prevent this manipulation.

A probabilistic approach is also susceptible to potential manipulation of estimates, with now additional estimates, namely of variances and correlations, being also input to the analysis. The variances of estimates can be manipulated to a certain extent to get an IRR closer to anything desired. As the uncertainty of an estimate decreases, the variance also decreases. As an extreme, considering no uncertainty reduces the probabilistic estimates to deterministic ones. For given expected values, results using a deterministic approach will be maintained by the probabilistic analysis under any variance assumptions. For implementation purposes, it is recommended that project proponents be required to use consistency in variances for analysis inputs for the without-CER case and the with-CER case.

How does consistency between without-CER and with-CER assumptions influence additionality and viability?

6.7 For the wind power project of Section 6.7, consider the sensitivity of the results to changes in the cash flow variance, intercomponent correlation and intertemporal correlation assumptions:

a. Cash flow variance. Consider Scenario 3. Let the cash flow deviations from the most likely value be equal, but with changing magnitudes.

b. Cash flow correlation. For intercomponent and intertemporal correlation, consider the two bounds – perfect positive correlation and independence.

6.8 For the multiproject case (Section 6.8), examine the appropriate choice of owner-selected feasibility levels and their meaning in terms of investment risk.

6.9 For the multiproject case (Section 6.8), how might the calculations be inverted, such that the project cash flow requirements and/or the number of projects could be determined and/or the project configurations could be determined that lead to desired feasibilities (rather than the other way around) or that lead to maximizing feasibility. This could follow a combined inferential and trial-and-error approach, or an optimization approach.

6.10 For the multiproject case (Section 6.8) assume that the investor has specified a required 0.75 feasibility level, that is, the investor is interested in the time by which a feasibility level of 0.75 is reached.

For the data given in Section 6.8 for separable projects, plot feasibility versus number of projects. Plot the incremental decline in months to feasibility with increasing number of projects. Show that the feasibility of the whole improves with the number of projects, and the decreasing/increasing relative contribution as more projects are added.

6.11 For all the results given in Section 6.8, construct a table showing the time taken to reach a feasibility level of 0.75 for each of the various project assumptions considered in the case example. Estimate the times to reach a level of 0.75 from the diagrams.

The calculations show the influence of different assumptions concerning individual and collective feasibility on overall feasibility. Would you anticipate that the trends indicated in this table to be repeated for other investment projects with similar cash flow patterns?

The calculations indicate that for the example, it is better to build in viability at the individual project level than at the combined project level, when separable investments are present. Is this a general result?

Chapter 7

Real options

7.1 INTRODUCTION

This chapter shows how the probabilistic background and formulation dealt with in the previous two chapters can be used to value options. This and the following chapter treat real options. This chapter looks at generic call- and put-style options, while Chapter 8 considers more detail, together with some applications. Chapter 9 treats related financial options.

Deterministic present worth is popularly used to establish the viability of an investment, including that involving options, commonly using a discount rate that is adjusted for risk. However it ignores the ability of the investor to make a choice during the lifespan of the investment and, as such, calculates a lower value of the overall investment when compared with real options analysis. Yet industry is hesitant to use real options analysis.

Part of this current hesitancy in using real options analysis derives from the requirement of having to use financial options as analogies, and then use financial options analysis methods – Black–Scholes or binomial lattices – and part from the unintuitive nature of the analysis; there is a need to establish a corresponding financial option and volatility, and this can raise difficulties. Having said this, there are also many people who are comfortable with existing real options analysis based on financial options thinking. While remedies have been suggested to resolve the various challenges of current real options analysis, there is no consensus, for example, on how to determine an analogous value for volatility as used in financial options. This points to the need for a method for real option estimation independent of any mention of matters specific to financial options, such as volatility. The approach given here does not require knowledge of financial options.

Within this context, this chapter gives a method for estimating the value of a real option, an approach which incorporates the familiarity and intuitive feel that people prefer in present worth (PW) analysis, while also evaluating future choices as preferred by people who advocate real options analysis. It encompasses the best of both worlds. It avoids the need to know or determine, for example volatility; the uncertainty is incorporated within the variance terms of the analysis input variables.

Let anything favourable to the investor be a cash inflow, and anything unfavourable be a cash outflow. That is, no distinction is made between buying and selling; rather each investment is interpreted from the viewpoint of the investor, not in any strict accounting sense. Based on this, it is shown that all options can be valued by using a single formula, here referred to as the Carmichael equation,

$$OV = \Phi \times \text{Mean of PW upside}$$

where Φ is $P[PW > 0]$ and the PW upside is the area of the PW distribution to the right of the origin ($PW > 0$). OV is an estimate of the option value. This applies for both call-style and put-style options. That is, knowing the distribution of PW, the option value can be calculated. There is no need to distinguish between the different real option types such as defer, expand or abandon. (In Chapter 9 it is also shown that, with the same thinking, there is no need to distinguish between the different types of financial options.) By using the Carmichael equation, all of the different option types, whether financial or real, can be put under a single framework.

In getting to the distribution for PW, this book adopts a second order moment approach, but Monte Carlo simulation could equally well be used if the investor prefers. The second order moment approach does not require any assumptions to be made on the distributions of the variables of cash flows, interest rates and so on. Each potential investment is reduced to its respective cash flows, and analyzed based on these. A spreadsheet is all that is needed to perform the calculations. At the end of the calculations, a distribution is fitted to the calculated present worth expected value and variance, and this distribution can be whatever the investor thinks is most appropriate, for example a normal distribution.

The approach given offers an investor a way of estimating the value of a real option, alternative to using financial options analogies. Values obtained are similar to those calculated by the Black–Scholes equation (abbreviated to 'Black–Scholes' here), suggesting equivalence between the two. However the approach given offers several advantages over Black–Scholes including that uncertainty in the exercise value can be incorporated, the exercise can occur at multiple points in time, interest rates and variances specific to different positive and negative cash flows are allowed, and it is more intuitively appealing and straightforward.

Black–Scholes gives a range of real option values dependent on the value for volatility used. Each way proposed in the real options literature for calculating volatility gives a different number. The resolution of this to everyone's satisfaction may never occur because of the inapplicability of volatility in a real options sense. There is no one definitive approach for calculating the volatility for real options.

A comparison of Black–Scholes with the book's approach is carried out in two ways (see the appendix in this chapter). First, a structural comparison shows that each captures the value of a real option in an equivalent way. Differences are noted. Second, the analysis of numerous option cases illustrates that the approach given leads to similar results to Black–Scholes over a wide range of values for the analysis inputs. The comparison suggests that the approach can be used as a substitute for Black–Scholes for both real options and financial options.

7.2 A PRESENT WORTH FOCUS

The form of Black–Scholes (Equations A7.1 and A7.2, which apply to a call option) suggests that it may be possible to calculate directly the value of an option using familiar present worth analysis, rather than by using assumptions that underpin Black–Scholes, and use the result as an estimate of the option value.

Consider how a result equivalent to Equation (A7.2) can be obtained by using a probabilistic present worth focus, with very few assumptions.

Consider a call option involving an exercise cost K at time T, from which follows a stream of net favourable asset cash flows (at and after T). The situation at time T, the time of exercising the option is shown in Figure 7.1, where the net favourable asset cash flows have been discounted back to time T. Let the subscript T denote values at time T.

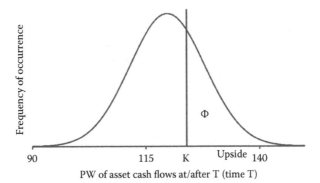

Figure 7.1 Example situation at time T. PW_T – (present worth) distribution of asset cash flows (K treated separately) occurring at or after time T, discounted to time T, showing upside of investment. The exercise cost K corresponds to where the vertical axis is located; present worth to the right of the vertical axis is referred to as the upside. (From Carmichael, D. G. et al., *The Engineering Economist*, 56(4), 295–320, 2011.)

Values to the right of K correspond to worthwhile investments. This is referred to here as the upside, and reflects the value embedded within the investment due to uncertainty.

The present worth relative to time T is,

$$PW_T = (PW \text{ of asset cash flows at/after } T)_T - \text{Exercise cost}$$

Define the option value at time T, OV_T, as the expected value of PW_T, evaluated on the assumption that the investment is only made if it is worthwhile. Then,

$$OV_T = E^*[(PW \text{ of asset cash flows at/after } T)_T - \text{Exercise cost}]$$

To denote that the investment is only made if it is worthwhile, the symbol * is used.

This can now be discounted to time 0. The option value at time 0,

$$OV = pwf \times OV_T$$

$$= E^*[pwf \times (PW \text{ of asset cash flows at/after } T)_T - pwf \times \text{Exercise cost}]$$

$$= \text{Expected* PW of asset cash flows at/after T}$$
$$- \text{Expected* PW of exercise cost} \tag{7.1}$$

OV is evaluated assuming that the investment is only made if it is worthwhile.

That is, C in Black–Scholes, Equation (A7.2), and OV in Equation (7.1) are obtained through expressions that are structurally the same. In both cases, the assumption is that the investment will only occur if it is worthwhile. OV becomes an estimate of C.

A distinction between the exercise cost, K, and the subsequent asset cash flows was made in deriving Equation (7.1) in order to make a comparison with Black–Scholes and for no other reason. Using the present worth approach, there is no need to make this distinction. As well, the approach allows K to be probabilistic; this is contrasted with Black–Scholes, which assumes a deterministic K. In the present worth approach, the exercise cost is treated as but another cash flow. Accordingly, assuming that the investment is only made if it is worthwhile, that is investments with negative present worth are not made (probability of zero),

$$OV = E^*[PW]$$

$$= (1 - \Phi) \times 0 + \Phi \times \text{Mean of PW upside}$$

$$= \Phi \times \text{Mean of PW upside} \tag{7.2}$$

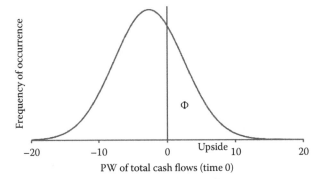

Figure 7.2 Example situation at time 0. Present worth distribution of total cash flows. Present worth to the right of the vertical axis (positive present worth) is referred to as the upside; the area under the curve to the right of the vertical axis is Φ; the mean of the upside is measured from PW = 0. (From Carmichael, D. G. et al., *The Engineering Economist*, 56(4), 295–320, 2011.)

where PW is the present worth of all cash flows at time 0, and Φ is the investment feasibility. Equation (7.2) is referred to here as the Carmichael equation.

The mean of the present worth upside is measured from PW = 0. The situation at time 0 is shown in Figure 7.2.

As drawn in Figure 7.2, the mean or expected value of the total present worth distribution is less than 0, implying that the deterministic present worth of the investment is less than zero, and hence under conventional deterministic thinking, the investment would not be undertaken. However, in general, the mean of the total present worth distribution can lie anywhere along the horizontal axis. (Where an investment has E[PW] positive (Φ > 0.5), an analysis might not be done of the option, because it would be a viable investment irrespective of any option value.)

7.3 CALL AND PUT OPTIONS

OV (Equation 7.2) is an estimate of an option's value. It is the upside potential of the investment.

Let anything favourable to the investor be a cash inflow, and anything unfavourable be a cash outflow. That is, no distinction is made between buying and selling; rather each investment is interpreted from the viewpoint of the investor, not in any strict accounting sense. A call option has an exercise value as a cash outflow, and the asset value (discounted cash flows resulting from exercising the option) as a net cash inflow. A put option has the reverse of this. A call option is exercised only if the asset value (discounted cash flows resulting from exercising the option) exceeds the exercise value. A put option is exercised only if the exercise value exceeds the asset value (discounted cash flows resulting from exercising the option). Both call-style and put-style options can be evaluated from OV provided the signs of the relevant cash flows are taken into account.

Using the subscript C to denote a call option and P to denote a put option, then,

$$OV_C - OV_P = E[PW] \qquad (7.3)$$

If the distribution of PW for a call option is drawn, then the call option is working with the part of the PW distribution to the right of the origin, while the put option is working with the part of the PW distribution to the left of the origin (but evaluated with the signs reversed).

7.4 IMPLEMENTATION

There is no need to distinguish between different real option types mentioned in the literature, such as defer, expand or abandon. Each is reduced to its respective cash flows.

The appraisal method outlined in Section 5.6, incorporating the Carmichael equation, is based on cash flows involved in the investment. Even the exercise value is considered to be just another cash flow (positive or negative as the case may be depending on the type of option) in the year(s) that exercising takes place. For example, expansion implies a negative cash flow (cash flow out); selling implies a positive cash flow (cash flow in). Only cash flows connected directly with exercising an option need estimating. Any amount spent now – the premium – in order that the future option is possible, is not included in the analysis. If the time to exercising is probabilistic, then an estimate of the probabilities of exercising at different time periods is required.

The discounting of cash flows to PW follows Chapters 6 and 10. The characteristics of PW can be obtained from E[PW] and Var[PW]. This in turn is used to calculate the option value, OV, using the Carmichael equation (see Chapter 5). To calculate the mean of the present worth upside, a formula based on the equation for the distribution adopted, for example a normal distribution, can be used. Alternatively, for an approximate value, the upside part of the present worth distribution can be divided into vertical strips and its area and centroid calculated as a structural engineer would calculate for a member cross section. See Chapter 5.

For a given Var[PW], a larger E[PW] means higher feasibility and higher option value, while a lower E[PW] means a lower feasibility and lower option value. That is, for a given Var[PW], as Φ increases/decreases, so too does OV increase/decrease respectively. Accordingly the preferred investment, where alternatives exist, is that with the largest OV (dependent on the cost of the premium or option right). With an individual investment, what is considered a minimum acceptable option value will depend on other circumstances and intangibles surrounding the investment, and the cost of the option right. It is noted that there will always be an option value because of the positive tail of the distribution representing PW. However, the option value will approach zero for investments with low feasibility. And so investors might like to make a decision based on both the option value, OV, in conjunction with the feasibility, Φ.

What feasibility, Φ, and the option value, OV, are telling is that all investments may turn out as losses and also turn out as gains.

7.5 EXAMPLE CALCULATION OF OPTION VALUE

Let the exercise cost be $125 000 at year 5, and assume that it can generate expected positive cash flows of $25 000 for each of years 5 to 10. Let the interest rate be 10% per annum. The deterministic present worth of such an investment is −$3250, implying that the investment is not worthwhile using such a measure. However variability produces an upside.

1. Estimates are required of the expected values and variances of each of the positive cash flows. Assume that the investor estimates that optimistic and pessimistic values for these cash flows are ±50% either side of the expected values, which are here assumed to be also most likely values. Using the Chapter 5 expressions for calculating expected values and variances from optimistic, most likely and pessimistic estimates, the expected value and variance of each yearly positive cash flow are $25 000 and 17 360 ($²) respectively.

2. Discounting the exercise cost (negative) and positive cash flows to the present, assuming independent cash flows,

$$E[PW] = -\$3250 \text{ (as in the deterministic case)}$$

$$Var[PW] = 26\,280 \times 10^3 \ (\$^2)$$

3. Calculate the feasibility Φ, the probability that the present worth is positive. This is the area under the positive part of the present worth distribution. Calculate the mean of the present worth upside. These calculations can be done using either the equations for the distribution assumed for present worth (here a normal distribution is used) to get accurate values, or for an approximate value, divide the upside part of the present worth distribution into vertical strips and calculate

its area and centroid as a structural engineer would calculate for a member cross section. See Chapter 5. Here $\Phi = 0.27$, and upside mean = \$3110, giving, from Equation (7.2),

$$OV = \$830$$

This is an estimate of the option value. (With a volatility of 3.1% derived using Equation (A7.3), Black–Scholes gives an option value of \$1420. As a proportion of the exercise cost of \$125 000, the difference is 0.47%.)

APPENDIX: RELATIONSHIP TO BLACK–SCHOLES

A7.1 Introduction

The approach given is a substitute for Black–Scholes. It offers investors a useful tool for valuing real options. The two do equivalent calculations. Through a comparison of their structures, it is seen that each approach captures the value of a real option in an equivalent way. Numerous numerical examples show that both result in essentially the same option values.

The approach given offers a number of advantages over Black–Scholes: uncertainty in the exercise value can be incorporated, interest rates and variances specific to different cash flows are allowed, various combinations of cash flows and cash flow correlations (complete independence through to complete correlation) are allowed, and it is more intuitively appealing and straightforward.

The numerical examples presented for comparison purposes here give small differences (generally less than 1% of the exercise value) between the given approach and Black–Scholes for a range of values of analysis inputs. Accuracies way less than this occur in the input cash flow estimates in any capital investment. Hence the approach given should be acceptable in most investment decisions. As well, finance is but one of many issues, including intangibles, involved in capital investment decision making.

As Black–Scholes has been subject to adjustments and refinements over the four decades since its introduction, so too it is envisaged that adjustments and refinements will be carried out on the approach given, as investors come to understand its strengths, weaknesses and idiosyncrasies.

A7.2 Black–Scholes equation

Consider a call option. (Put options can be similarly considered.) Black–Scholes values a European call option according to,

$$C = N(d_1)S_0 - N(d_2)Ke^{-rT} \tag{A7.1}$$

where

$$d_1 = \frac{\ln\left(\dfrac{S_0}{K}\right) + (r + 0.5\sigma^2)T}{\sigma\sqrt{T}}$$

$$d_2 = d_1 - \sigma\sqrt{T}$$

and,

C	the value of the call option
$N(d_i)$	the cumulative standard normal distribution of the variable d_i
S_0	the current asset value
r	the risk-free rate of return; Black–Scholes uses continuous time discounting
K	exercise price or value
T	the option life; the (expiry) time available to exercise the option
σ	volatility

Although Black–Scholes is straightforward to use if values for the analysis inputs can be estimated, its derivation and explanation are more complicated.

$N(d_1)S_0$ is the expected (deterministic) present worth of the asset value if the option finishes in the money, that is, if the asset value is above the exercise value at T. $N(d_1)$ is the risk-adjusted probability that the asset value will finish above the exercise value at T (Nielsen, 1992).

$N(d_2)Ke^{-rT}$ is the expected (deterministic) present worth of the exercise value if the option finishes in the money. $N(d_2)$ is the risk-adjusted probability that the option will be exercised. Ke^{-rT} is the present worth of the exercise value (Nielsen, 1992).

For an exercise value K, Black–Scholes calculates the discounted payoff that would occur if the current asset value increases above K, adjusted for the probability that the asset value will be greater than the exercise value.

That is, Black–Scholes, Equation (A7.1), in simplified notation effectively calculates the option value as,

$$C = \text{Expected* PW of asset value at T} \\ - \text{Expected* PW of exercise value} \qquad \text{(A7.2)}$$

assuming that the investment is only made if it is worthwhile; the symbol * is used to denote this assumption.

A7.3 Criticisms of Black–Scholes for real options

Possible reasons for a reluctance to adopt real options analysis, via financial options analogies, include difficulty in determining values for the analysis inputs for Black–Scholes, acceptance of deterministic discounted cash flow analysis, and lack of understanding of real options analysis.

Real options analysis does not seem currently popular among industry professionals; it has been suggested that real options analysis might overestimate the uncertainty of real investments, based on the fact that options analysis is defined for financial options that have their own bases and assumptions.

The use of Black–Scholes to determine the value of a real option is questioned in the literature for a number of reasons. Besides the high level of mathematical sophistication required of users to understand its derivation and meaning, the following issues are raised in regard to using financial option tools when determining the value of a real option:

- The arbitrage principle is difficult to accept for real options since real assets are not freely traded. The arbitrage principle is a key assumption in financial option valuation. A number of authors suggest being cautious when using financial options theory when the assets cannot be freely traded.
- While geometric Brownian motion may be a suitable model for fluctuating underlying asset price movements, it has no applicability to real assets.
- The difficulty of determining the volatility in the real options case is discussed by many writers. There is also the concern that volatility may change during the long life of investments involving real options, and this contradicts the assumption that the volatility for the life of the option can be determined and is constant. While volatility has relevant meaning for fluctuating underlying assets, it has no transferable meaning for real options. The total variability in a real investment cannot be captured by exercising the option, because variability still lies in the future. There is also criticism of many of the ways used for establishing volatility in the real options case.
- Financial options are usually exercised instantaneously. This is not the case for many real options. There are also issues associated with delaying investment, and establishing single dates of exercising.
- The decision relating to a financial option cannot change the value of the underlying asset, while the decision relating to a real option can, that is the cash flows for a real asset can change with exercising a (real) option.
- Black–Scholes uses continuous time discounting, whereas most real investment calculations are based on discrete time discounting (Section 5.2.3).

The challenges to real options are discussed at length in the literature and whilst many solutions have been suggested, there is no common agreement on how to resolve all the issues listed above. The approach given in this book addresses this.

The basis of the approach given is no more than a conversion of any investment into a collection of cash flows characterized by their expected values and variances, and examining the present worth of these cash flows. As such the method requires no assumptions on arbitrage or volatility, and has no need for geometric Brownian motion assumptions. It also permits whatever cash flows, and uncertainties attached to these cash flows and other analysis inputs, that the investor wishes to incorporate. Further comment is made on these characteristics in the following sections.

A7.4 Comparison of the approach given with Black–Scholes

It is shown above that C in Black–Scholes, Equations (A7.1), and OV in Equation (7.2) are obtained through expressions that are structurally the same, and OV becomes an estimate of C. Further comparison is made here in terms of

- Differences between Black–Scholes and the approach given
- A range of numerical example cases

All numerical example cases examined demonstrate that the approach given determines values that are essentially the same as Black–Scholes.

A7.4.1 Differences with Black–Scholes

There are some differences between the approach given and Black–Scholes. These are discussed here.

Distribution of asset value. Black–Scholes assumes that the underlying asset value is lognormally distributed. A lognormal distribution has a left-hand tail bounded by zero. This means that the asset value can never be less than zero. Asset values are generally not able to follow any distribution that permits negative values. For example, it is not possible to have stock prices that are negative. However, there is nothing preventing discounted cash flows (and present worth), which represent the underlying asset value in a real investment, from taking negative values. Distributions that allow this include the normal distribution.

Hence, the assumption of a lognormal distribution for the underlying asset value (discounted cash flows resulting from exercising the option) may be unrealistic for real assets. A normal distribution or other distributions that allow negative values might be considered to be better representations for real asset values (discounted cash flows resulting from exercising the option). The approach given allows any distribution for present worth but presumes that most people will adopt a normal distribution.

Volatility. Although both the approach given and Black–Scholes incorporate variability or uncertainty, they do it in different ways. Black–Scholes incorporates uncertainty through volatility; the approach given incorporates uncertainty through variances of the cash flows and other analysis input variables. The literature lists numerous issues involved in estimating volatilities for real options cases, including where negative cash flows exist, or similar comparison real investments don't exist. The variance calculation under the approach given however is straightforward for both positive and negative cash flows, other uncertain analysis input variables, and for one-off investments. In both the approach given and Black–Scholes, an increase in variability or uncertainty causes an increase in the value of the option and vice versa; the higher the uncertainty of the cash flows and other analysis input variables, the higher the value of the option.

Black–Scholes captures the total variability through volatility on exercise. However in real options, variability still lies beyond the date of exercise; the approach given acknowledges such future variability.

Black–Scholes uses a constant volatility. The approach given can accommodate different variances for different cash flows and other analysis input variables for each time period.

Exercise value. Black–Scholes assumes that the exercise value K is deterministic, that is, it has no uncertainty. The approach given allows for probabilistic K, and hence is more general. Uncertainty in the exercise value can be accommodated, if the investor desires. To convert to a deterministic K, the variance of K is set to zero.

Black–Scholes assumes that the exercise value occurs at one point in time. The approach given can assume any general exercise form including a series of amounts, or amounts distributed over time.

Interest rate. Black–Scholes uses a constant risk-free rate. The approach given can accommodate a different interest rate for each cash flow component (positive or negative), but generally assumes that an interest rate, which is the cost of capital for the investor or the opportunity cost of capital or other, but unadjusted for any uncertainty, will be used because the uncertainty in the cash flows and other analysis input variables, and the distance into the future is accounted for in the variance terms for the analysis input variables.

Black–Scholes uses a constant rate. The approach given can vary the rate at each time period, and incorporate probabilistic interest rates (Chapter 10).

Expiration time. While an increase in the time until expiry (T) in Black–Scholes causes an increase in the value of the option because of a higher level of uncertainty, it is not as obvious how an increase in T affects the calculations in the approach given, because all cash flows are discounted similarly. Increased uncertainty with time within the approach given is reflected in the larger estimated variances associated with the cash flows and other analysis input variables as time extends into the future.

Continuous and discrete time. Black–Scholes uses continuous time discounting. Although the approach given is in terms of discrete time discounting, it can accommodate both continuous and discrete time discounting. However it is remarked that most real options applications would be developed in terms of discrete time, whether this is days, weeks, months, years or longer, and calculations would be done on spreadsheets on this basis.

Cash flows. Any combination of positive and negative cash flows is possible under the approach given. Positive and negative cash flows can be series or individual sums at various points in time.

Negative cash flow values are possible, in contradistinction to the Black–Scholes or binomial lattice methods. And multiple sources or causes contributing to asset cash flows and values (the so-called rainbow option) are treated no differently to a single source or cause, using the approach given.

Option type. With the approach given, just as there is no need to distinguish the exercise amount from other cash flows, there is no need to distinguish all the different types of real options; all that is necessary is to know the cash flows for any particular investment.

A7.4.2 Numerical comparison

Figures 7.3 to 7.6 show indicative results of numerical studies on real options. Differences (C – OV) between Black–Scholes and the approach given are expressed as a percentage of the exercise value in order to remove the influence of different magnitudes of exercise amounts. Interest rates,

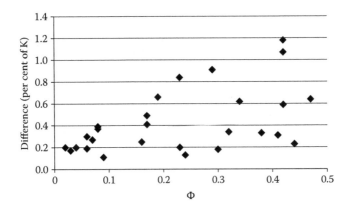

Figure 7.3 Difference (per cent of K) versus feasibility, Φ. (From Carmichael, D. G. et al., *The Engineering Economist*, 56(4), 295–320, 2011.)

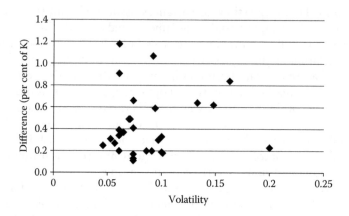

Figure 7.4 Difference (per cent of K) versus volatility, σ. (From Carmichael, D. G. et al., *The Engineering Economist*, 56(4), 295–320, 2011.)

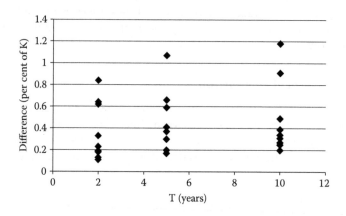

Figure 7.5 Difference (per cent of K) versus time to expiration, T. (From Carmichael, D. G. et al., *The Engineering Economist*, 56(4), 295–320, 2011.)

exercise times, asset cash flows and variability/uncertainty are changed in different combinations to examine their impact. This leads to numerical studies covering a range of volatilities, feasibilities and option values. Options close to the money and far from the money are included.

Data common to all results reported in Figures 7.3 to 7.6 are as follows:

Option: to expand (call option), with an exercise cost at T followed by positive cash flow
Variance of exercise cost: 0, in order to compare with Black–Scholes
Present worth: normal distribution assumed

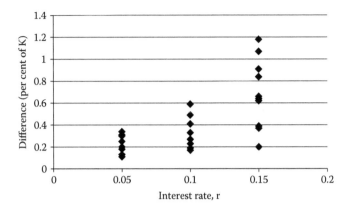

Figure 7.6 Difference (per cent of K) versus interest rate, r. (From Carmichael, D. G. et al., *The Engineering Economist*, 56(4), 295–320, 2011.)

Volatility is calculated from (Hull, 2002),

$$\sigma = \sqrt{\frac{\ln\left(\dfrac{Var[S_T]}{E[S_T]^2} + 1\right)}{T}} \qquad (A7.3)$$

where $E[S_T]$ is set equal to the expected present worth at T of the (positive) cash flows occurring at or after T, and $Var[S_T]$ is set equal to the variance of the present worth at time T of the (positive) cash flows occurring at or after T; S_0 is set equal to the expected present worth at time 0 of these (positive) cash flows.

Note that the differences shown in Figures 7.3 to 7.6 are not additive but are the same differences represented against Φ and analysis input variables in turn.

It is appreciated that there are numerous ways for calculating volatility, and different volatilities will give different option values when calculated using Black–Scholes. There appears to be no agreement on what is the correct volatility to use for real options, and hence there will be no agreement on what is the correct Black–Scholes option value. In such circumstances the approach given has nothing exact to compare with.

Consider now the above example but with a large negative cash flow at the end of the asset's lifespan. Let r = 0.10, T = 5 years, positive cash flow from year 5, and the negative cash flow in year 11 be approximately twice year T positive cash flow. Figure 7.7 plots the difference (C – OV as a percentage of K) for the case with positive cash flow only, and the case with a negative cash flow at the end of the asset's life. Because of the additional cash flows, and hence additional cash flow variance, introduced in the latter case, the associated volatility is higher.

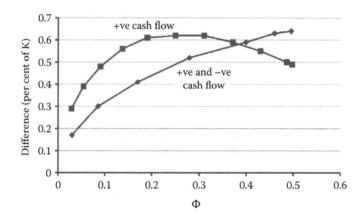

Figure 7.7 Difference (per cent of K) versus feasibility, Φ; r = 0.10; T = 5; positive cash flow from years 5 to 10; negative cash flow in year 11. (From Carmichael, D. G. et al., *The Engineering Economist*, 56(4), 295–320, 2011.)

A7.5 Advantages of the approach given

The advantages of the approach given over Black–Scholes may be summarized as follows:

- The approach given is more directed at real options. Black–Scholes was developed for calculating the value of financial options and is based on stock market practices and information. To use Black–Scholes, information about a capital investment needs to be converted into financial option terms. In contrast to this, the approach given uses variables relevant to capital investors, such as cash flow expected values and variances.
- The approach given is an intuitive tool for estimating the value of real options. Users are able to see the probability distribution of the present worth of their investment and then calculate the option value. The use of Black–Scholes is based on showing that the real option case is analogous to a financial option case and as such can be valued in the same manner. Black–Scholes appears complicated to many people and does not offer an investor an intuitive understanding of how the value of the real option is being calculated. While investors may understand why real options analysis values particular investments higher than deterministic present worth, investors are not able to see how Black–Scholes calculates the extra value.

The approach given offers investors a substitute for Black–Scholes, one that resolves the main reasons for its lack of popularity in real options analysis.

A listing of advantages of the approach given over Black–Scholes when applied to real options is as follows:

- The approach given is more in line with conventional capital investment calculations, which historically have preferred deterministic present worth analysis.
- The approach given is intuitively appealing. Users are able to visualize an investment's upside.
- The distribution describing the present worth may be chosen to suit the situation, but it is presumed that a normal distribution will be used, because of its support and ease of use, and it also allows for negative values.
- There is no need to estimate volatility, a concept not applicable to real assets, and hence indefinite in its estimation. The uncertainty incorporated through volatility in Black–Scholes is incorporated instead through the variance terms. Increasing volatility and variances both increase the option value. Future variability (beyond the exercise date) is acknowledged.
- Correlation between cash flows is allowed. The approach given can cater for complete independence through to complete correlation of cash flows. Correlation leads to larger present worth variances and larger option values.
- The exercise value may be probabilistic, and exercising can occur at a single point in time or at different points in time.
- Different interest rates can be used for different cash flows, but generally the approach given assumes that an interest rate, which is the cost of capital for the investor or the opportunity cost of capital or other, but unadjusted for any uncertainty, will be used because the uncertainty in the cash flows and the distance into the future is accounted for in the variance terms. Probabilistic interest rates can be used (Chapter 10).
- Both discrete time and continuous time discounting versions are possible.
- Any combination of cash flows, including negative cash flows, is possible.
- There is no need to distinguish between the various real options types. All that is required is the relevant cash flows.

Exercises

7.1 Assume some suitable numerical values and test the accuracy of the approach given when compared with Black–Scholes.

7.2 How do users of financial options methods applied to real options cases argue about the applicability of arbitrage, volatility and geometric Brownian motion?

7.3 What does volatility mean in a real options sense? Or does it have no meaning?

7.4 Where an investment has E[PW] positive ($\Phi > 0.5$), an analysis might not be done of the option, because it would be a viable investment irrespective of any option value. Investments with Φ close to 1 should not need an options calculation to justify viability. Investments with Φ close to 0 would need an options calculation to support viability. At what value of Φ between 0 and 0.5 does the transfer occur for an options calculation – from being needed to not being needed?

7.5 Increasing the variance of the cash flows and other analysis input variables increases the variance of the present worth, which in turn increases the option value. Comment on the potential for abuse by investors wanting to justify an investment by unrealistically assuming large uncertainty, which equates to estimating large cash flow variances.

7.6 Few users of real options tools would have the mathematical background to understand the basis of the derivation of the Black–Scholes equation. Comment on the dangers of using Black–Scholes as a black/grey box.

7.7 Chapter 10 introduces uncertain interest rates. The approach given applies to any assumptions on interest rates. How would you anticipate option values to change with changing uncertainty in interest rates?

7.8 Would you anticipate the difference between the approach given and Black–Scholes to increase with increasing feasibility, Φ, and interest rate, r?

7.9 Why does Black–Scholes, when applied to real options, have trouble dealing with negative cash flows?

7.10 How sensitive would you anticipate real option values to be with different assumptions on cash flow correlation?

Chapter 8

Real option types and examples

8.1 INTRODUCTION

Chapter 7 outlined a useful approach for valuing real options based on conventional investment thinking using cash flows. It was shown that options can be valued by using a single formula, the Carmichael equation,

$$OV = \Phi \times \text{Mean of PW upside}$$

in conjunction with a second order moment analysis (or Monte Carlo simulation if desired). Here Φ is $P[PW > 0]$ and the PW upside is the area of the PW distribution to the right of the origin ($PW > 0$). OV is an estimate of the option value. This applies for both call-style and put-style options. It assumes that anything favourable to the investor be classed as a cash inflow and anything unfavourable as a cash outflow. That is, no distinction is made between buying and selling; rather each investment is interpreted from the viewpoint of the investor, not in any strict accounting sense.

The approach given leads to the conclusion that all option types can be reduced to the same thing.

The approach given also avoids the need to draw an analogy with financial options (in order to use financial options tools), and consequently avoids all the attendant drawbacks associated with this analogy. Using financial options tools is a forced analogy, where there is a requirement to find the equivalent of a distinct premium, stock price and exercise price, as occurs in stock options. That is, there is a requirement to force a framework on real options, a framework that is not compatible with real options.

The appendix in Chapter 7 compared the given approach with Black–Scholes, and for this comparison purpose it was necessary to put both the given approach and Black–Scholes under a similar structure. However, with confidence that the given approach leads to results consistent with Black–Scholes, it is possible to get rid of that structure and look purely at using the approach to value a given investment. That is, the stock options structure is no longer necessary for real options valuation purposes. A real option can be looked at similarly to any general investment, namely in terms of cash flows. It does not matter what the cash flows are branded as (for example, branded as an exercise amount); they are just treated as positive or negative cash flows as the case may be.

Real options analysis allows for prudent capital budgeting decisions to be made for projects during their evaluation phases and investments generally. Real options analysis puts a value on *flexibility* or *adaptability*. Typically, money (equivalently, a premium) is spent now in order to have flexibility or choice in the future.

This chapter explores some of the many possibilities with real options.

8.2 CATEGORIZATION

Real option types can largely be reduced to the following broad categories:

- Option to *expand* (equivalently, to buy, invest or similar – a call option)
- Option to *contract* (equivalently, to sell, divest or similar – a put option)

However, using the approach of Chapter 7, there is no need to even make this distinction. All that is necessary is to look at the cash inflow and cash outflow of the investment, as would be done in any conventional nonoptions investment.

The option to contract includes the option to *abandon (termination)*. Within and connected to these option categories, there may be *delay* or *deferment* (of the time to exercising), *rainbows* (multiple sources of variability), the ability to *choose* between option types (also called *switching*), the ability to *change* to something else (this includes changing asset *output mix*, asset *input mix* and asset *operation scale*), and other options done in combination (for example in a *sequential* or *staged* fashion, or more generally referred to as *compound* options). However again, using the approach of Chapter 7, there is no need to even make these distinctions that historically have built up in the real options literature (as a direct result of adopting financial options analogies). All that is necessary is to look at the cash inflow and cash outflow of the investment, as would be done in any conventional nonoptions investment.

The applications of real options are spread across many industry sectors. The jargon may differ across industries, but provided the cash flows are identified, the analysis is the same.

Usual treatments of real options categorize them into call (buy) and put (sell) options, and involve quite clever thinking in being able to interpret a real option within the restrictive (and inapplicable) framework of financial options theory. In essence, this involves identifying some deterministic proxy that can be called an exercise value, and a random variable proxy that can be called an asset value. In some cases this may not be possible, and so simplifying assumptions are necessary, or only restrictive cases can be dealt with.

All of this style of thinking can be avoided if instead attention is focused on the cash flows involved in any investment. This, in all cases, then looks at whether cash coming in exceeds that going out, as in nonoptions investment analysis, in order to evaluate the worth of having future flexibility, and hence be able to compare with any cost outlaid now (equivalent to a premium in financial options theory).

The different real options that appear in the literature are first described here. They are then summarized in terms of their cash flows.

Option to expand. At some future time there arises the possibility to invest further, and this further investment desirably yields net positive cash flows. The expansion will be undertaken only if the future resulting net positive cash flows are greater than the cost of expanding.

An example application is where there are two possible sequential project investments where the second project is possible only if the first project is carried out. The first project may be regarded nonviable on its own, with a low Φ value. The option value of the second project may convert the investment in the first project to one of being more viable.

The option to expand is analogous to the call option case in financial options. A future sum (exercise cost) is paid to expand an asset/facility and this provides an ongoing (beyond T) return. At time T, the equivalent of the financial options asset price is the discounted (to time T) future cash flows resulting from exercising the option.

Option to contract. Downsizing operations may or may not yield a benefit. If, at the time of exercising, there is a future benefit (the difference between with and without contraction is favourable), then the option will be exercised.

Abandonment (option to abandon) is a special case of contraction. Here the asset has a one-off salvage (liquidation, residual) value. This is compared with the value in continuing operations. The abandonment option will be exercised if, at the time of exercising, the future cash flows are less than the salvage value.

For example, in a supply agreement, a company may have to pay a premium to suppliers in order that the suppliers are flexible in the quantity agreed to supply. This provides the company with the option to contract or abandon in terms of product received within the agreement.

Delay. The worth of an investment may change over time. How such worth changes over time reflects the particular investment, and there is no one general formulation; the cash flow modelling for each particular investment is chosen to suit that investment. The approach adopted to incorporate the influence of delay will depend on the situation at hand. Irrespective of how the cash flows change over time, delay influence is readily incorporated into the standard option types – expand and contract. That is, no special analysis is required for delay; all that is necessary is to identify any altered cash flow values and times of occurrence.

In a nondividend paying financial option, it is never optimal to exercise early due to the inherent time value of the option which will be lost if early exercise occurs. However, in a real options context, if the exercise of the option is delayed and the associated cash flows are not delayed into the future, then there will be foregone cash flows. If the cash flows are delayed into the future, then their present worth will be less because they are discounted over a longer period. Under such a situation, it may be optimal to exercise early to avoid delay influences. The particular influence of a delay will affect the value of the option.

Delay influences may lead to cash flows (resulting from exercising the option) being:

- Lost or no longer available
- Delayed but not lost
- Reduced or changed

The influence of delay may be important in a real option. For financial options, the time to exercising is typically relatively short, and the influence of delay may have no or little consequence. However, for real options, ignoring the influence of delay might result in over- or undervaluation.

Early exercising of an option (an American option in financial options terminology) or just generally having a variable exercising time can be handled as follows. Let the time to exercising the option follow a probability distribution. Let p_t be the probability that the exercise takes place in period t, t = 0, 1, 2, ..., T, and let $PW_{(t)}$ be the associated present worth. Then,

$$E[PW] = \sum_{t=0}^{T} p_t E[PW_{(t)}]$$

$$Var[PW] = E[PW^2] - \{E[PW]\}^2 \tag{8.1}$$

The choice of the probability distribution for the exercising time is up to the user, and would reflect the investor's belief in the likelihood of exercising within the time T. Example possible distributions are uniform where exercising is equally likely at any time up to T; a declining triangular distribution where the value of the option is perceived as decreasing with time; or a rising triangular distribution where exercising is more likely the closer time gets to T. The case $p_T = 1$ and $p_t = 0$, t = 0, 1, 2, ..., T–1 is the European option in financial options terminology.

Rainbows. In financial options terminology, and considering the simplest option type, the value of the option is due to the variability in the underlying asset value. However, in real options, often there may be multiple sources or causes of variability associated with whether the option will be exercised or not. Using financial options theory, this introduces multiple forms of volatility, and hence complications are introduced.

Using the approach of Chapter 7, the treatment is unchanged from that of standard options. For example, consider a company with the rights to extract a natural resource (oil, iron ore, etc.). In this example, there are at least two sources of uncertainty: the price of the resource in the future and the quantity of the resource available to be extracted. Both variables will affect the cash flows of the project and therefore influence whether the company will go ahead with extracting the resource. The value of the resource is calculated directly from resource price multiplied by resource quantity, that is, the product of two random variables, which is readily handled if the random variables are expressed in terms of their expected values, variances and covariance (or correlation). For this example, the correlation coefficient between resource price and resource quantity might be assumed to be zero.

Choose. At some point in the future there may be the possibility of choosing between expansion, contraction and continuing operations unchanged. For example, a manufacturing facility, at some point in the future may be able to expand existing capacity to meet new demand through additional investment; abandon operations and recover the salvage value of the facility; contract capacity (for a certain period or indefinitely) should demand drop, by reducing the scope of operations; or continue existing operations.

At exercise time, the most attractive option is chosen. The value of having a choice between options can be established from the value of each option separately. The option to choose represents a high degree of flexibility in an investment.

Where there is both the potential to expand and the potential to contract, these will be mutually exclusive because it is not possible to simultaneously expand and contract. Therefore it would appear reasonable that the larger of the two option values would be chosen. Should one of the two options be available for exercising earlier than the other, at this earlier exercising date the two options can be reevaluated based on updated cash flows, and the more appropriate one chosen at this date.

Summary. Table 8.1 summarizes the cash flows for each real option type and is given in terms of a base case and a changed case. Both the base case and the changed case refer to the situation at and following the exercise date of the option. Exercising the option brings about the change. Not exercising the option leaves the base case in place. The option is exercised if the changed case is better than the base case.

Financial options theory makes a lot of fuss about the distinction between call and put options. For real options, using the approach of Chapter 7,

Table 8.1 Options cash flow summary

	Base case (at and postexercise date)		Changed case (at and postexercise date)	
Option type	Positive cash flows	Negative cash flows	Positive cash flows	Negative cash flows
Financial call	nil	nil	Asset/stock price/value, S_T	Exercise price/value, K
Financial put	nil	nil	Exercise price/ value, K	Asset/stock price/value, S_T
Real expand	nil	nil	Resulting +ve cash flows	The cost of expansion; resulting −ve cash flows
Real contract	Existing +ve cash flows	Existing −ve cash flows	Resulting +ve cash flows; salvage value	Resulting −ve cash flows; nil

the distinction is not important; all that is necessary is to look at the cash inflow and cash outflow of the investment itself, as would be done in any conventional nonoptions investment. Assume that anything favourable to the investor is classed as a cash inflow and anything unfavourable as a cash outflow. That is, no distinction is made between buying and selling; rather each investment is interpreted from the viewpoint of the investor, not in any strict accounting sense. Doing this, all options can be analyzed with a common framework. There is no need to make a distinction between expand, contract and other options.

8.3 OUTLINE DEMONSTRATIONS

The examples given here illustrate real options analysis on projects with future flexibility value resulting from uncertainty. With real options analysis, such projects have increased viability compared with deterministic present worth analysis. Without real options analysis, such projects might be rejected on the basis that the deterministic analysis produces a negative present worth.

8.3.1 Option to expand

A company is considering expanding into a new market in 5 years' time (T = 5). The cost to expand (negative cash flow) is estimated as X_5 and will result in estimated net positive cash flows in the following 5 years of X_i, i = 6, 7, ..., 10. The interest rate is r% per annum.

Assume that these X_i, i = 5, 6, ..., 10 are most likely values and also expected values. (Other assumptions can be made.) For deterministic calculations, the present worth is, from Chapter 6 (appropriately accounting for signs),

$$E[PW] = \sum_{i=5}^{10} \frac{E[X_i]}{(1+r)^i}$$

This will establish whether the expansion is viable or not according to deterministic thinking, depending on the sign of E[PW].

Consider now the probabilistic calculations. With most likely values in hand for the cash flows, it remains to estimate pessimistic and optimistic values for the cash flows. From Chapter 5, this will lead to $E[X_i]$ and $Var[X_i]$, i = 5, 6, ..., 10.

With suitable assumptions on the correlations between the cash flows (see Chapter 5), then Chapter 6 gives,

$$E[PW] = \sum_{i=5}^{10} \frac{E[X_i]}{(1+r)^i}$$

$$Var[PW] = \sum_{i=5}^{10} \frac{Var[X_i]}{(1+r)^{2i}} + 2\sum_{i=5}^{9} \sum_{j=i+1}^{10} \frac{\rho_{ij}\sqrt{Var[X_i]}\sqrt{Var[X_j]}}{(1+r)^{i+j}}$$

Using the 'method of moments' from Chapter 5 and assuming a normal distribution for PW (Chapter 5), then the probability distribution for PW is completely defined.

Chapter 5 gives how the feasibility, Φ, and mean of PW upside are now calculated. The option value is then,

$$OV = \Phi \times \text{Mean of PW upside}$$

This establishes how much the uncertainty in the future cash flows provides in value along with the ability to make a decision in the future should the cash flows be favourable. The company has to reconcile the option value with whatever the cost is now (equivalently, the premium) of providing for future expand flexibility.

8.3.2 Option to contract/abandon

Consider an investment (negative cash flow) now in equipment worth X_0, with a lifespan of 10 years. During these 10 years, it is estimated to generate net positive cash flows of Y_{il}, i = 1, 2, ..., 10. Should the equipment not prove popular, at the end of year 5 (after receiving any positive cash flow for that year) the contractual buyout value (positive cash flow) by another is Y_{52}, and results in Y_{il}, i = 6, 7, ..., 10 being foregone (equivalently a negative change in cash flow). The interest rate is r% per annum. Consider from buyout onwards.

Consider first the investment without buyout. This gives, $X_i = Y_{il}$, i = 6, 7, ..., 10. Assume that these X_i, i = 0, 6, 7, ..., 10 are most likely values and also expected values. (Other assumptions can be made.) And (appropriately accounting for signs), from Chapter 6,

$$E[PW] = \sum_{i=6}^{10} \frac{E[X_i]}{(1+r)^i}$$

Along with i = 0, 1, ..., 5 cash flows, this will establish whether the investment is viable or not according to deterministic thinking, depending on the sign of E[PW].

Now consider the deterministic case with buyout included as definitely happening. The investment only goes over 5 years. This gives $X_5 = Y_{52}$. And,

$$E[PW] = \frac{E[X_5]}{(1+r)^5}$$

Along with i = 0, 1, ..., 5 cash flows, this will establish whether the investment is viable or not according to deterministic thinking, depending on the sign of E[PW].

For the possible (not definite) buyout included case, assume now that the cash flow estimates contain uncertainty. With most likely values in hand, it remains to estimate pessimistic and optimistic estimates for the cash flows. From Chapter 5, this will lead to expected values and variances of the relevant X_i, Y_{il} and Y_{i2}. If buyout occurs, Y_{il}, i = 6, 7, ..., 10 are foregone (equivalently negative cash flows). For evaluating the flexibility of the buyout option, the cash flows are $X_5 = Y_{52}$, and $X_i = -Y_{il}$ (explicitly incorporating the sign), i = 6, 7, 8, 9, 10. With suitable assumptions on the correlations between the cash flows, from Chapter 6,

$$E[PW] = \sum_{i=5}^{10} \frac{E[X_i]}{(1+r)^i}$$

$$Var[PW] = \sum_{i=5}^{10} \frac{Var[X_i]}{(1+r)^{2i}} + 2\sum_{i=5}^{9}\sum_{j=i+1}^{10} \frac{\rho_{ij}\sqrt{Var[X_i]}\sqrt{Var[X_j]}}{(1+r)^{i+j}}$$

Using the method of moments from Chapter 5 and assuming a normal distribution for PW (Chapter 5), then the probability distribution for PW is completely defined.

Chapter 5 gives how the feasibility, Φ, and mean of PW upside are now calculated. The option value is then,

OV = Φ × Mean of PW upside

This establishes how much the uncertainty in the future cash flows provides in value along with the ability to make a decision in the future should the cash flows be favourable. The company has to reconcile the option value with whatever the cost is now (equivalently, the premium) of providing for future buyout flexibility.

The case of partial contraction, rather than full contraction as covered here, is dealt with in the same way.

8.3.3 Delay

To evaluate when is the best time to exercise an option, enumeration can be used. To evaluate the best time to exercise an option, a range of exercise times are selected. For each exercise time in turn, the cash flows are adjusted to reflect the influence of delay, and the option value is calculated. This process is repeated for all exercise times, and that which gives the largest option value is the optimal time to exercise.

8.3.4 Sequential options

Sequential options, that is, where exercising one option is dependent on having exercised an earlier option (or, a chain of options), are perhaps best thought of in a reverse time sense. That is, the last option is valued first, followed by the second last, third last, etc., continuing to the first option. The collective option can be valued from the cash flows and option values, going backwards in time, and making sure to take into account the signs of the cash flows and types of options.

Bellman's principle of optimality stated in the 1950s is relevant here (Carmichael, 1981, p. 139; 2013):

> An optimal policy has the property that whatever the initial state and initial decision are, the remaining decisions must constitute an optimal policy with regard to the state resulting from the first decision.

An example is the phasing or staging of projects. At each phase, there is an option to continue with the project (equivalent to an option to expand) or an option to contract/abandon the project. Exercising each option must be completed before the next option is available.

Parallel options, where there is a dependent option and an independent option, can be thought of similarly. The dependent option is valued first, followed by the independent option. The collective option can be valued from the cash flows and dependent option value, going in the order of independence, and making sure to take into account the signs of the cash flows and types of options.

8.3.5 Comparing options

When comparing alternative options, awareness has to be exercised in dealing with time frames and differing scales, as with the deterministic present worth measure.

8.4 CLIMATE CHANGE AND INFRASTRUCTURE

With climate change comes increasing uncertainty. With increased uncertainty comes the need for rethinking appraisal studies of infrastructure investment. Flexible or adaptable infrastructure offers the potential of better financial viability over existing nonflexible, nonadaptable infrastructure because of this uncertainty. Real options offer a sound methodology for not only acknowledging future uncertainty but also allowing investors to value flexibility or adaptability.

With climate change, deterministic steady state assumptions are not applicable. Infrastructure, in the future due to climate change, could be anticipated to have to deal with increased temperatures, altered rainfall patterns, altered frequencies of extreme weather events and sea level rise. These in turn will lead, for example, to changed demand patterns, increased maintenance and operation costs, decreased longevity, increased costs of retrofitting, changed land use and demographics, and more frequent disruption to use. And all of these consequences are known only to within defined probabilities. Future costs and benefits for infrastructure are inherently probabilistic, and any investment appraisal must necessarily take into account the uncertainty. Deterministic appraisals can no longer be justified.

With climate change, the future is uncertain because of the incomplete science and associated lack of confidence in prediction. Future benefits and costs and infrastructure lifespans cannot be predicted with any degree of accuracy, because of uncertainty associated with anticipated sea level and temperature rises, and changed occurrences and magnitudes of extreme weather events and rainfall patterns. Increases are anticipated in infrastructure maintenance, repair and operation costs, damage, and insurance premiums, while demands on infrastructure, energy, water and transport will change – both increase and decrease on a situation-by-situation basis. The locations of infrastructure needs will also change as demand changes. The operation of infrastructure will be disrupted more frequently. The longevity of infrastructure will decrease and external facades of infrastructure will experience accelerated degradation. Infrastructure will need to be replaced more frequently. Much has been written on this.

Climate change will expose vulnerabilities in existing infrastructure and infrastructure established along business-as-usual lines. And such vulnerability could be anticipated to vary between locations. Existing infrastructure could be anticipated to have limited ability or capacity to adapt, and may be found to be inadequate. Infrastructure intended for long-term use may prove inadequate.

> There is considerable uncertainty about the timing and intensity of future climate change especially at regional and local scales. The economic situation contributes further uncertainty. Nonetheless decisions about developments, many irreversible, will continue to be made, and these need to increasingly take into account an awareness of both climate change and uncertainty about its specific local implications. One sensible approach for large investments is to undertake staged developments that allow for future expansion or additional adaptive features to be implemented contingent on certain climatic thresholds being surpassed (NCCARF, 2009, pp. 14–15).

Three main alternatives for new infrastructure are possible:

   I. Design for present-day conditions and abandon/replace in the future because of climate change ('failure'), whereby the longevity of the infrastructure is restricted.
  II. Design with flexibility or adaptability such that the infrastructure can be adapted/modified/upgraded in response to the climate, and the infrastructure is tailored to the changing climate or adapts to the changing climate.
 III. Design for possible future conditions whereby the infrastructure is overdesigned/excessive at present but adequate for the longer term.

Each alternative represents different levels of feasibility. Options analysis allows the evaluation of the viability of having flexible or adaptable infrastructure, such that rational investment decisions can be made within the uncertainty introduced by climate change.

Where existing infrastructure is being refurbished/retrofitted/renovated, the same alternatives apply.

The cash flow diagrams for each of the three cases I, II and III might follow something like Figure 8.1.

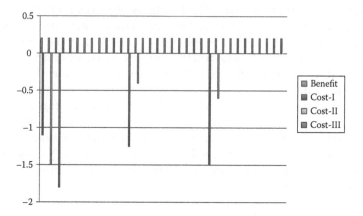

Figure 8.1 Schematic example cash flow diagrams for the three possible infrastructure investment cases. Benefits are above the line. The costs below the line are in the order I, II and III. (From Carmichael, D. G. and Balatbat, M. C. A., The Incorporation of Uncertainty Associated with Climate Change into Infrastructure Investment Appraisal, Conference – Managing Projects, Programs and Ventures in Times of Uncertainty and Disruptive Change, Sydney, Australia, 2009.)

Infrastructure alternatives I, II and III can be incorporated under the treatments in Chapters 5 to 10. Each alternative can be represented by a different set of cash flows over time. The nature, magnitude and sign of these cash flows will differ between the three alternatives.

Example

To demonstrate the types of results obtained, consider three investments of the form shown in Figure 8.1. The numbers and correlation type assumed are not important; rather it is the methodology that is important. Uncertainty (variances) in the benefits and costs is assumed to increase with time. Assumptions on infrastructure costs, benefits and lifespans, and interest rate are made. Benefits are assumed correlated. Costs are assumed correlated. Benefits and costs are assumed uncorrelated.

Figure 8.2 shows example present worth distributions for cases I, II and III. Feasibility for each is the area under the curve to the right of the origin (upside). Each alternative demonstrates different feasibilities. The option value for II is obtained from its present worth distribution.

Building in flexibility to infrastructure may take many forms including defending, modifying (retrofitting, alteration), retreating, relocating and abandoning.

The likely impacts of climate change need to be recognized and an adaptive management approach to designing and managing investment is essential.

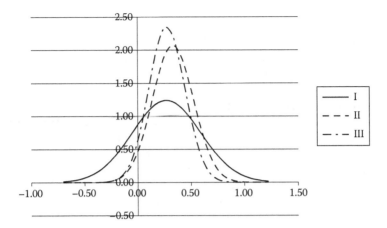

Figure 8.2 Example present worth density functions for infrastructure alternatives I, II and III. (From Carmichael, D. G. and Balatbat, M. C. A., The Incorporation of Uncertainty Associated with Climate Change into Infrastructure Investment Appraisal, Conference – Managing Projects, Programs and Ventures in Times of Uncertainty and Disruptive Change, Sydney, Australia, 2009.)

Climate conditions will change considerably over the life of long-lived infrastructure ... The capacity for such assets to incorporate adaptation treatments or adjustments to their maintenance regime will in part determine their resilience to accelerated degradation of materials and fatigue of structures due to increased intensity and frequency of extreme events (storms, wind, rainfall, bush fire). ... Integrating ... renewal options in long-lived infrastructure would also help enable periodic improvements to these assets as knowledge improves (NCCARF, 2009, p. 23).

8.5 ADAPTABLE, FLEXIBLE INFRASTRUCTURE

Incorporating flexibility in infrastructure allows for the exploitation of possible future opportunities not existing at the present time. Flexibility can also allow for staged upgrades so that capital is spent (delayed capital expenditure) only when it is needed, rather than committing the full expense up front, and can even offer the opportunity to scale back in order to cut costs, thus reducing future financial exposure. As such it might be regarded as a risk management tool. Flexibility has a particularly important place in relation to uncertainties in the future climate and fast-track projects. Being able to adapt to future needs and situations keeps infrastructure relevant in an ever-changing world.

Examples of in-built flexibility in infrastructure are numerous; however heretofore missing from the literature was a method of valuing this flexibility that is consistent across all infrastructure cases and incorporates the uncertainty of the future. Since flexibility comes at a cost, it is important to know what the returns will be, including whether the returns will exceed the costs. Accordingly, being able to establish the value that flexibility adds to infrastructure would be useful. The approach of Chapter 7 permits this.

The time to execute the available choice can be set as fixed (European option) or variable (American option). Generally, the option value increases with uncertainty (larger variance). That is, the greater the uncertainty of the cash flows into the future, the greater the value of the incorporated flexibility. It may not be that incorporating flexibility is the best answer in all cases. There is nothing in the analysis to suggest any possible generalization. Each infrastructure project will require its own analysis. The flexible alternative will then have to be compared with the other alternatives available. Generally, where the investment benefits are close to the investment costs (feasibility Φ close to 0.5), or where feasibility has a value less than 0.5, having flexibility will add value to the investment and may provide that increment which converts a marginal project into a justifiable one. And generally where the benefits exceed greatly the costs (feasibility Φ close to 1), having flexibility will not be necessary in order to justify the project.

Infrastructure, over its lifetime, can be anticipated to be subject to the changing forces of nature because of uncertainties in the future climate. Infrastructure, such as buildings, can also be subject to use (purpose) and ownership changes. Accordingly, infrastructure may have to reinvent itself a number of times over its lifetime. Flexibility is all about leaving decisions open because of future uncertainty. Having flexibility or adaptability would thus be considered useful. Fast-tracking of projects almost always requires in-built flexibility to cope with uncertain future design.

The notion of having flexibility or adaptability in infrastructure is not new. Some infrastructure has possibilities for utilizing deliberate flexibility (the basis of this section), while some infrastructure ends up having fortuitous flexibility. For example, the Sydney Harbour Bridge is said to have been designed such that a second, lower deck could be added after completion; the Eiffel Tower was saved from demolition through its usefulness for communications; the Kuala Lumpur Stormwater Management and Road Tunnel does as its name implies, namely doubles as a road and a flood drain; the open-space design of Heathrow Airport Terminal 5 allows for future changes and reconfigurations if needed with minimal disruptions to operations; while the Colosseum in Rome is said to have had the ability to host a wide variety of events, including gladiatorial fights, animal hunts, naval battles, chariot races, mythological plays and public executions. The incorporation of flexibility in infrastructure design can require ingenuity. Its implementation may also raise issues. Nevertheless, many interesting examples exist all over the world.

Here, the term *flexibility* is used in the real options sense that a deliberate decision and initial investment has been made to allow for future possible changes to infrastructure, and such changes may or may not occur depending on future uncertain circumstances; the term flexibility is not used to mean redundancy in a system reliability sense, or as in the fortuitous flexibility case.

Having flexibility can act as a hedge against future uncertainties.

Currently there is no preferred method of how such flexibility is valued. The approach of Chapter 7 offers a way forward. Versatility and adaptability can be desirable and advantageous, depending on the situation. A premium is paid now in return for having choices in the future; such choices are available but may not be acted on, depending on the future situation.

8.5.1 Regional water supply

Consider the case example of water supply to a regional community, where the usage and needs depend on future population growth. A regional town was undergoing relatively slow population growth but there had been a sudden surge in recent housing development applications, which could lead to a large population increase in the coming years. This led to an examination of the future water supply for the town. One possibility, amongst several that were considered, included that of having storage that could be upgraded should the town's population and its water usage undergo an upsurge. The question arises as to the value of having adaptable capacity, given the possibility that the extra capacity may never be used; this value can then be compared with the additional capital cost.

Data for the study included estimates/forecasts of the town's population for the next 25 years under various growth scenarios. Income derives from the sale of water, which was assumed proportional to the population. Cost estimates included those for maintenance, pumping and capital expenditure. Expected values and variances of cash flows were obtained from optimistic, most likely and pessimistic estimates.

In the calculations, the ongoing costs and income, Y_{ik}, were assumed independent (the correlation coefficients $\rho_{k\ell}$ are 0), while net cash flows over time were assumed correlated (the correlation coefficients ρ_{ij} are 1). This leads to an expected value and variance, respectively, of the present worth of the extra capacity. For an assumed interest rate, the option value is calculated. This is the value of extra capacity flexibility, which can be compared with the additional capital cost.

8.5.2 Regional road upgrade

Consider the case example of a road, where preparatory activities now allow the possibility of a future upgrade. A regional arterial road was experiencing increased traffic, with attendant increasing delays and accidents.

Population growth was anticipated to increase. At some point an upgrade of the road might be needed. However, before an upgrade was possible, property acquisition would be required, and this needed to be done now, before land rezoning occurred, if it was going to happen. The question arises as to the value of the acquisition, in terms of the future flexibility it allows, and whether it is sensible to outlay costs now in preparation for the possible upgrade.

Estimates of future benefits and costs contain uncertainty. Data for the study included population estimates/forecasts, and estimates of the cost of the upgrade and ongoing maintenance costs (optimistic, most likely and pessimistic). Benefits (reduced travel time, congestion, fuel use and accidents) were converted to equivalent monetary values. Estimates were made in terms of optimistic, most likely and pessimistic values for 20 years into the future; from these, the expected values and variances were obtained of the benefits and the costs.

For calculation purposes, the upgrade was assumed to occur in 10 years, but other assumptions could be used, or this time could also be considered a random variable (American option). Similar assumptions on correlation or independence and interest rate were made as for the water supply example. The Y_{ik} are the costs and benefits in each year. $\rho_{k\ell} = 0$, $\rho_{ij} = 1$. This gives E[PW] and Var[PW], from which the option value is obtained. This is the upgrade flexibility value, which can be compared to the capital cost of the necessary acquisition.

A related options road example involves undertaking, now, additional excavation and subbase work to allow for the possibility of expanding (beyond those currently being constructed) the number of lanes in the future, should future demand require additional capacity.

8.6 CARBON FARMING INITIATIVE

The Carbon Farming Initiative (CFI), based on Australian legislation, is a voluntary carbon offset scheme, which provides landowners with an incentive to reduce carbon emissions. Carbon credits are issued for every tonne of carbon emissions mitigated or sequestered. These credits can then be sold on an appropriate carbon market.

CFI was introduced in 2012 to counter the carbon emissions being produced by the agricultural sector, to reduce pollution, manage impacts of climate change, protect Australia's natural environment and improve farm productivity.

Within CFI, a sequestration project (for example, reforestation and revegetation) presents an opportunity to make a gain through possible termination. Broadly, since the price of carbon varies over time, if future carbon prices are less than present prices, there is the potential to terminate the sequestration and gain income in other ways (for example, by selling the product which might be timber, by selling the land, or by changed land use). Along the way, there is also the benefit of carbon credits (and their compound interest) issued prior to termination. On termination, the carbon credits are surrendered at the price applying at the date of termination.

Consider that termination occurs in year T, provided it is worthwhile. The landholder doesn't terminate the project if it isn't worthwhile. That is, it is a 'right' rather than an 'obligation' to terminate. It is an option to terminate (abandon, contract). The landholder earns income (made up of carbon credits, minus retention money and costs) each year i = 1, 2, ..., T – 1 up to termination. With termination, the landholder pays back credits worth $K = Y_{T2}$ (using the carbon price at year T) at year T, but gains Y_{T1} through sale of the product, or changed land use. X_T is comprised of Y_{T1} and Y_{T2}, and all are random variables.

Provided K is less than Y_{T1} and the income (made up of carbon credits, minus retention money and costs – all discounted to year T) that will be foregone in years after T, then termination is worthwhile.

This formulation is now the same as that for any real option, and the option value, or the value of having the option to terminate is OV.

Rather than T being fixed, the equivalent of an American option can be also analyzed, where T is a random variable. Another way of approaching this is to evaluate the option for increasing T values, and see how the option value changes with life of the sequestration project.

Because there is the possibility of terminating a sequestration project and making money out of the termination, it could be argued that this is counter to the intent of the CFI idea, which is to help reduce emissions. If the cash flows are attractive enough, investors will terminate their projects, thereby releasing carbon back into the atmosphere. Investors will have made money, while atmospheric carbon will not have been reduced. Alternatively, the option value could be viewed as an extra carrot to get people to invest in CFI, with the hope that termination won't occur.

8.7 MINING OPTIONS

A mining project has characteristics generally different to other investment projects. A typical project involves large capital outlays, and the returns may not start until after a number of years. The lifetime of a mine may be long and the returns subject to the uncertainties beneath the ground and fluctuating market prices.

Uncertainties in any mining investment analysis result from the ore being mined (reserves and grade), the mining process, the cost of extraction, commodity prices, exchange rates, and political, social, governmental and environmental issues.

Issues surrounding mine opening and closing depend on ore price, operating costs and reserves. There may be flexibility to open and close mines, including the delaying of these. Options analysis allows the valuing of having this flexibility. Because mining projects contain considerable uncertainty, so the value of having this flexibility is high.

For example, consider the option of opening/closing a gold mine at time T. This is equivalent to an expand/contract option. While running, mine income will derive from the sale of ore (which value depends on the gold price, gold grade, recovery and unit cost of production – all random variables estimated from historical and geological information). There will be an initial capital investment as well as ongoing operation and maintenance costs over the life of the mine. For analysis purposes, ore grade and mill recovery might be regarded as being positively correlated, while gold price from year to year as being highly correlated. At certain gold prices it will be worthwhile to open/close the mine, both of which would involve expenditure. Calculating the value of the option to open/close the mine might be viewed as putting a value on managerial flexibility. Management is able to react as the gold price changes, or new technology emerges. With time, the uncertainty in the variables reduces.

The expand/contract option is evaluated according to the approach of Chapter 7, and gives the value OV.

Similar analyses may be done for other mining options.

8.8 FAST-TRACKED PROJECTS

A particular situation where flexibility and uncertainty is relevant is that of fast-track projects. Fast-tracking implies an overlapping of project phases, typically the overlapping of the design and construction phases; in order to reduce the project duration, the construction starts before the design is complete, however this is at the expense of certainty. The early design necessarily has to be such as to not restrict or constrain choices in the later design; that is the early design (and what is constructed based on this) needs to incorporate flexibility in order to accommodate many potential design possibilities in the later design.

The advantages and disadvantages, and costs and benefits, of fast-tracking are well documented. Typical fast-tracking costs are as follows:

- Extra initial direct construction and design costs
- Extra managerial costs dealing with design and construction simultaneously, and inefficiencies in the way the work is organized and resourced; possible increased rework

Typical fast-tracking benefits are as follows:

- Reduced project completion date (which may or may not have a monetary value, depending on the project)
- Value that results from the designer leaving the final decision on design to a later point in time (design flexibility); more designs explored; reduced potential for design errors and changes/variations

Currently no method is available to evaluate the flexibility value offered by fast-tracking. The approach of Chapter 7 provides such a method.

Consider a building project. To have flexibility, the premium could be the cost (labour, materials) expended in additional foundation work together with extra managerial costs, and indirect costs resulting from disrupted scheduling of the work. In return, delaying final design decisions gives an opportunity to develop designs more fully and examine design alternatives, which in turn may generate benefits. There may be opportunities to react as the project progresses by downsizing, expanding or changing. This flexibility offers value.

8.9 CONVERTIBLE CONTRACTS

The term 'convertible contract' is an umbrella term that describes a flexible type of contract, which starts out in one form and then later converts to another form.

Convertible contracts have been used in the finance industry – in financial bonds, securities and venture capital transactions. They have also been used in insurance policies and adjustable rate mortgages, allowing the holder of the financial product to change contractual terms, such as the length of insurance or the type of interest rate applicable.

Convertible contracts also have a place in project work, related for example to infrastructure. Typically at the start of a project, the work is ill-defined and hence remuneration to the contractor might best be on a cost-reimbursable (cost plus, prime cost) basis. As the project progresses, the work becomes better defined, and hence remuneration to the contractor might best be on a fixed price (lump sum, schedule of rates, unit price, guaranteed maximum price) basis (Carmichael, 2000).

By incorporating flexibility, specifically the option to convert from one contractor remuneration form to another, convertible contracts allow for projects to commence under one contract form and later convert to another. This flexibility allows for the adjustment of risk throughout the project as new information becomes available.

Convertible contracts allow the parties to agree on one set of terms initially and different terms later depending on certain circumstances or the outcomes of specific events. The terms are considered to be 'converted'. A way of understanding what takes place is by considering two separate contracts; effectively, the old contract containing the initial terms is converted to the newer contract containing the converted terms. In the context of, say, construction contracts, these terms may relate to a range of things, including payment, the conditions of work, the programming/scheduling of work or the scope (of work).

It is unlikely that all the terms of a contract are subject to change, however. The terms that are flexible would be agreed upon by the parties and clearly identified as terms that may be later converted. The extent to which these terms may be converted would also be explicitly stated, and are essentially contingent on the agreement between the parties. Alternatively, contracting parties may choose to agree on the extent of conversion upon initial engagement, without the need for later amendments or the introduction of new terms.

The contractual power to change might be conferred on both or either party through a 'conversion clause', but more likely on the owner in an owner-contractor relationship. Such a clause allows the parties the right, but not the obligation, to convert according to the details within the clause.

There are a number of commercial drivers that might encourage parties to adopt such contracts. From the owner's perspective, benefits such as hedging against potential financial losses and increasing contractor value (related to output and cost) exist. For the contractor, advantages arise from the opportunity to profit from the conversion in some cases and to show good faith, thereby increasing its potential for future business.

The commercial viability of convertible contracts is dependent on the cost of the option. The approach of Chapter 7 provides a simple tool to value options associated with convertible contracts.

The legal perspective of convertible contracts needs to address issues such as termination of the first contract, formation and enforceability of the second contract, and novation.

Exercises

8.1   In Section 8.3, option to expand, let the cost to expand be $1M, which results in estimated net positive cash flows in the following 5 years of $0.22M each year. Use an interest rate of 5% per annum.

For the deterministic calculations, what is the present worth of the cash flows? Is the expansion viable according to deterministic thinking?

Consider the probabilistic calculations for this example. Let the estimated pessimistic and optimistic values for the cash flows be ±50% of the mean. Assume perfect correlation between the net positive cash flows. Calculate Var[PW], Φ, mean of PW upside and the option value OV.

Does the option value convert a nonviable investment into a viable one?

8.2 In Section 8.3, option to abandon, let the initial investment in equipment be $500M, and this generates net positive cash flows of $80M for each of the 10 years. Should the equipment not prove popular, in 5 years the contractual buyout value by another is $400M. Use an interest rate of 5% per annum.

Consider first the investment without the buyout. What is the deterministic present worth of the investment over the 10-year lifespan? Is the investment viable according to deterministic thinking?

With definite buyout, what is the deterministic present worth of the investment?

Does variability in the cash flows convert the investment into a viable one? Consider the value of the buyout option. Assume the buyout value at year 5 is deterministic. Assume that the cash flow estimates for subsequent years contain uncertainty. Let the pessimistic and optimistic estimates for the cash flows be ±50% of the mean. Assume perfect correlation between the cash flows. Calculate E[PW], Var[PW], Φ, mean of PW upside and the option value OV.

Does the option value convert a nonviable investment into a viable one?

8.3 Option to contract. Consider a company with two similar plants. A potential competitor may undercut this company's product, and so the company is considering scaling back its operations through divesting itself of one of its plants. The savings from doing this are estimated to be $200M. It is anticipated that this may occur in year 5. With two plants, the annual income is $70M. With one plant, the annual income is $35M. Use an interest rate of 5% per annum.

For the deterministic calculations, using the above as most likely and also expected values, what is the present worth of both plants over 10 years?

Now consider the value of the option to contract. The calculations are little different to the option to abandon. The estimated savings are $200M at year 5, and the cash flow estimates are assumed to contain uncertainty. Let the pessimistic and optimistic estimates for all the cash flows be ±50% of the mean. Assume perfect correlation between the cash flows. Calculate E[PW], Var[PW], Φ, mean of PW upside and the option value OV.

Does the option value convert a viable investment into a more viable one?

8.4 Does the answer to the optimal time to delay an option always lie on a constraint in the optimization formulation? (For terminology, see Carmichael, 2013.)

8.5 Based on the water supply example of Section 8.5, delineate what you believe are the relevant cash flows in the evaluation of the option, and delineate the method you would use for calculating the option value.

8.6 Based on the road upgrade example of Section 8.5, delineate what you believe are the relevant cash flows in the evaluation of the option, and delineate the method you would use for calculating the option value.

8.7 For the Carbon Farming Initiative calculation of the value of flexibility in terminating, does the option value increase/decrease with T, the time to exercising the option?

8.8 For the case of an option involved in a convertible contract, outline what you believe are the cash flows involved in an option to convert.

8.9 Suggest a procedure by which sequential options might be evaluated. Would starting with the last option and working backwards in time facilitate this evaluation?

8.10 Assume that you were comparing the option value on two projects. One has an exercise date in 1 year, the other has an exercise date in 10 years. Are these two options directly comparable?

8.11 Consider the provision of spare parts and the reliability of plant or equipment. Purchasing the spare parts might be considered the premium paid for the possibility of repairing the equipment in the future should it break down. Equipment survivability might follow an exponential curve, and since the time, T, of exercising the use of the spares is unknown, the situation might be thought to be equivalent to an American option form. With no spares, the cost at T is the downtime cost. With spares, the gain might be the negative of the downtime cost. The exercise cost is zero. Is what has been described here a potential use of options analysis, or is it not an options situation because if you have the spares and breakdown occurs, you will always use the spares?

8.12 Consider the valuation of the adaptable design of sea walls. A sea wall that is constructed today has extra preparatory work done so that it can accommodate an upgrade should it be necessary in the future. Future sea level rises are uncertain, and hence the sea wall would be upgraded only should it be necessary. There is the ability, but not the obligation, to upgrade in the future should it be necessary. The premium is the extra cost involved now so that future upgrades can occur. The exercise cost is the cost of the upgrade. (1) Is this a similar situation to the spare parts case, in that it is not an options situation because if the sea level does rise, then upgrading will automatically occur? Or, (2) if you believe that it is a potential use of options analysis, what is the benefit of upgrading the sea wall, and hence what is the positive cash flow that is used in the options analysis?

8.13 Conversion between contract payment types. Commonly, projects start out with broadly defined information and this gets refined as the project progresses. Given the different characteristics and strengths and weaknesses of different contractor payment types, this implies that using one contract form for the entire duration of a project may not be optimal, and that a more prudent approach would be to tailor the payment type to the project stage, with transitions or conversions along the way. Information asymmetry, whereby the contractor is better informed of the work than the owner, and contractor self-interest leading to possible opportunistic behaviour, also hint at the need to tailor a contract to a project's situation in order that the owner's interests and the contractor's interests align. Decreasing uncertainty as a project progresses influences estimates and the cost and benefit of conversion, the timing of the conversion and the selection of the most appropriate payment type. Consider the particular conversion from prime cost to fixed price contracts within projects. How does this transfer risk between owner and contractor? What compensation would you anticipate that the owner might pay and the tender premium or fee that the contractor might include for having such a conversion inclusion? What associated practical implementation issues are there? Outline the framework of analyzing convertible contracts involving an option to convert, using the approach of Chapter 7.

Chapter 9

Financial options

9.1 INTRODUCTION

It is shown in Chapter 7 and below that all options, including all the various financial options, can be valued by using a single formula, here referred to as the Carmichael equation,

$$OV = \Phi \times \text{Mean of PW upside}$$

where Φ is $P[PW > 0]$ and the PW upside is the area of the PW distribution to the right of the origin ($PW > 0$). OV is an estimate of the option value. This applies for both call-style and put-style options. It assumes that anything favourable to the investor be classed as a cash inflow, and anything unfavourable as a cash outflow. That is, no distinction is made between buying and selling; rather each investment is interpreted from the viewpoint of the investor, not in any strict accounting sense. On knowing the distribution of PW, the option value can be calculated. The book's preferred way to get to PW is a second order moment analysis, but any alternative method can be used.

This chapter outlines the approach based on the Carmichael equation as an alternative to the Black–Scholes equation (abbreviated to 'Black–Scholes' here) and related methods. The approach given and Black–Scholes are shown to value the upside potential of options in equivalent ways, and give essentially the same results, but the approach given offers a number of advantages over Black–Scholes. The structural characteristics of the two methods are compared, with differences noted, and summary numerical comparisons are given for a large number of data sets. The approach given is readily understandable to those familiar with conventional deterministic PW analysis, is intuitive to apply, and requires a modest background in mathematics to derive. The method allows the extension beyond European options to American options, options with dividends, exotic options, carbon options and employee stock options.

Black–Scholes is used in options trading, for example for quoting and for establishing whether an options contract is over- or undervalued. It is straightforward to use but relies on a very sophisticated mathematical derivation using Ito stochastic calculus in conjunction with geometric Brownian motion, a background that most users do not have. To users, Black–Scholes thus might be viewed as a black box or at best a grey box.

On the other hand, most people are familiar with deterministic present worth (PW) analysis and feel comfortable using such an approach in capital investment calculations. The approach given for evaluating options outlined here uses a probabilistic extension of present worth, and hence the method has the straightforward familiarity and intuitive feel that people like in such an approach. The approach can work with or without knowing stock volatility, the one variable in Black–Scholes that is not observable, by incorporating any uncertainty within variance terms. A conversion between volatility and stock variance is available (Equations A9.3, A9.4).

A comparison of the approach given with Black–Scholes is undertaken in two ways. First, in terms of the underlying equations, it is shown that the approach given and Black–Scholes are structurally equivalent in the way they value the upside of options. At its core, options analysis focuses on the value embedded within investments due to uncertainty, and the flexibility in associated decisions. Differences in assumptions are noted. Second, extensive numerical testing over a range of volatilities, interest rates and times to expiration demonstrates that the approach given and Black–Scholes give essentially the same option values. It is accordingly argued that the approach given can be used as an alternative to Black–Scholes for option analysis, including nontraded options, options with dividends and American options.

All of the different option types can be most readily seen if PW is calculated via a second order moment analysis. Such analysis involves the usual mean or expected values of cash flows (that are used in the deterministic version of present worth), as well as cash flow variances, but not probability distributions. The mathematical background requirement is very mild. It does not require any assumptions to be made on the distribution of the stock price; it only requires an assumption to be made on the distribution of the present worth. A spreadsheet is all that is needed to perform the calculations.

9.2 A PRESENT WORTH FOCUS

The form of Equation (A9.2), based on Black–Scholes for a call option, suggests that it may be possible to calculate directly the value of an option using familiar present worth analysis, rather than by using assumptions that underpin Black–Scholes, and use the result as an estimate of the option value.

Consider how a result equivalent to Equation (A9.2) can be obtained by using a probabilistic present worth approach, with very few assumptions.

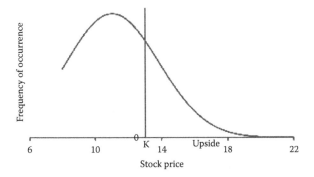

Figure 9.1 Example distribution of stock price, S_T. The vertical axis is located at the value of the exercise price K.

The situation at time T, the time of exercising the option is shown in Figure 9.1; the subscript T denotes values at time T.

Values to the right of K correspond to worthwhile investments. This upside reflects the value embedded within the investment due to uncertainty.

Define the option value at time T, OV_T, as the expected value of the net cash flow at T, evaluated on the assumption that the investment is only made if it is worthwhile (in the money, $S_T > K$). Then,

$$OV_T = E^*[\text{Net cash flow}_T] = E^*[S_T - K] \qquad (9.1)$$

To denote that the investment is only made if it is worthwhile (in the money), the symbol * is used. This can now be discounted to time 0. Using continuous time discounting, in order to compare with Black–Scholes, the option value at time 0,

$$OV = e^{-rT} OV_T = E^*[e^{-rT}S_T - e^{-rT}K]$$

$$= \text{Expected* PW of stock price} - \text{Expected* PW of exercise price}$$
$$(9.2)$$

OV is evaluated assuming that the investment is only made if it is worthwhile (in the money). In effect, OV is being calculated as in,

$$OV = e^{-rT} E[\max(S_T - K, 0)] \qquad (9.3)$$

For a lognormal assumption on the stock price and a risk-free rate, this leads to Black–Scholes; a proof of this is given for example in Hull (2006).

That is, C in Black–Scholes and OV above are obtained through expressions that are structurally the same. In both cases, the assumption is that the investment will only occur if it is worthwhile. OV becomes an estimate of C.

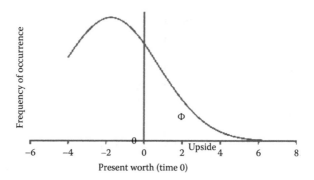

Figure 9.2 Example present worth distribution of $[S_T - K]$ at time 0. The term 'upside' is used here to denote positive present worth.

A distinction between the exercise price and the stock price was made in deriving Equation (9.2) in order to make a comparison with Black–Scholes and for no other reason. However in terms of getting to OV, there is no need to make this distinction; both the stock price and the exercise price are treated as cash flows. Accordingly, assuming that the investment is only made if it is worthwhile, that is an investment with a negative present worth is not made (probability of zero),

$$OV = E^*[PW]$$

$$= (1-\Phi) \times 0 + \Phi \times \text{Mean of PW upside}$$

$$= \Phi \times \text{Mean of PW upside} \tag{9.4}$$

where PW is the present worth of all cash flows at time 0, and Φ is the investment feasibility. Equation (9.4) is referred to here as the Carmichael equation.

The mean of the present worth upside is measured from PW = 0. The situation at time 0 is shown in Figure 9.2.

9.3 FURTHER COMMENT

Consider a call option. In line with Black–Scholes, at time T the stock, S_T, is a random variable and the exercise price, K, is deterministic. Their collective present worth (time 0), PW, is (using discrete time discounting),

$$PW = \frac{S_T - K}{(1+r)^T} \tag{9.5}$$

where r is the interest rate. The expected value and variance of the present worth become,

$$E[PW] = \frac{E[S_T] - K}{(1+r)^T} \tag{9.6}$$

$$Var[PW] = \frac{Var[S_T]}{(1+r)^{2T}} \tag{9.7}$$

Related sets of equations can be given for other option types.

The expected value and variance of S_T may be estimated or forecast in any way that the user wishes. Suggestions on how this might be done are given in Chapter 5. Expected values and variances may be estimated, additional to these suggestions, through forecasting based on past performance of the stock, or based on a time series of historical stock prices; using a proxy approach if the investor is aware of similar stocks; or if the stock volatility has been estimated, then the variance can be obtained from this through Equation (A9.4), and the expected value from the future worth of the current stock price. This relaxed way of estimating the stock price offers the capability of dealing with situations where stock price movements may not follow geometric Brownian motion as assumed in Black–Scholes, that is where stock price movements are atypical, for example where markets are responding to strong unusual external economic changes or crises; where companies are undergoing radical changes, reorganizations or performance fluctuations; and in dealing with nontraded options. The relaxation allows the investor to incorporate experience and arcane knowledge into the option calculation. Both Black–Scholes and the approach given share the need to estimate stock prices, and future stock prices are difficult to predict.

In both Black–Scholes and the approach given, the analysis variables are independent of risk preferences, and hence using the argument of Hull (2006), adopting a risk-free rate is appropriate. Both methods share the need to estimate a rate, and future rate movements are difficult to predict. The extension of the approach to include uncertain interest rates is given in Chapter 10.

An assumption is required on the resulting distribution of the present worth (but not on any distribution describing the stock price, which requires only expected value and variance estimates). Any distribution suitable as a descriptor of present worth can be used. Some users might prefer to adopt a normal distribution, because of its ease of use and familiarity. The shape of a normal distribution is completely defined on knowing its expected value and variance, and associated probabilities are readily evaluated.

Feasibility, Φ, is explained in Chapter 5. Having established E[PW] and Var[PW], Φ can be established, and Equation (9.4) is used to calculate OV, an estimate of the option value. It is the upside potential of the investment. Where competing investment choices exist, that with the largest option value might be preferred (dependent on the premium or cost of the option right). With an individual investment, what is considered a minimum acceptable option value will depend on other circumstances surrounding the investment, and the premium or cost of the option right.

Visualizing a normal distribution for PW (though this is not a requirement of the approach), because of the positive tail, there will always be an option value, even when the mean of the present worth distribution is negative (and there is also always a negative tail). This option value decreases as the value of Φ decreases, and approaches zero in the limit. Accordingly, investors might prefer to make a decision based on two numbers in conjunction: the option value, OV, and the feasibility, Φ.

In summary, the route to the option value is given in Section 5.6. This more generally incorporates probabilistic cash flows (exercise price and stock price) and probabilistic interest rate (Chapter 10). If the time to exercising is probabilistic (American option), then an estimate of the probabilities of exercising at different time periods is required.

9.4 EUROPEAN FINANCIAL OPTION

The European option is the case considered in Sections 9.2 and 9.3. Equations (9.6) and (9.7) are particular cases of equations given in Chapter 6 (discrete time discounting), where S_T and K are independent. In terms of the general notation and formulation of Chapter 6, this European option case can be set up as in the following. For a call option let

$$Y_{T1} = S_T$$

$$Y_{T2} = -K$$

$$Y_{Tk} = 0 \quad k = 3, 4, ..., m$$

$$X_i = 0 \quad i \neq T \tag{9.8}$$

S_T is the stock price (asset value), and K is the exercise price at time T. The exercise price K is taken as being deterministic, that is its variance is zero.

A put option reverses the signs of Y_{T1} and Y_{T2}. Put–call parity can be demonstrated.

9.5 EUROPEAN FINANCIAL OPTION WITH DIVIDENDS; TRANSACTION COSTS; TAXES

Dividends, transaction costs and taxes can be accommodated with both the European and American option cases, by regarding these as additional cash flows, and they may be probabilistic. If part of the option, they are incorporated directly into the option calculation; if not part of the option, they are incorporated into the overall investment viability calculation.

Using the notation and formulation in Chapter 6, if dividends can occur at any period i, i = 0, 1, 2, ..., n, then set the appropriate X_i to these anticipated dividends.

This formulation allows for uncertainty in the value of dividends. Should dividends be known with certainty, then their variances are set to zero. This is contrasted with usual approaches for the incorporation of dividends, which assume for analytical tractability that dividends are riskless (Hull, 2006).

Transaction costs and taxes are similarly considered as just additional cash flows. When taxes occur over time, and their magnitude will depend on the taxation laws specific to the user's locality, and the user.

9.6 AMERICAN FINANCIAL OPTION (CALL AND PUT)

Let the time to exercising the option follow a probability distribution. The choice of this distribution is up to the user and would reflect the investor's belief in the likelihood of exercising within the time T; uniform or triangular distributions might be suitable. Let p_t be the probability that the exercise takes place in period t, t = 0, 1, 2, ..., T, and let $PW_{(t)}$ be the associated present worth. Then,

$$E[PW] = \sum_{t=0}^{T} p_t E[PW_{(t)}]$$

$$Var[PW] = E[PW^2] - \{E[PW]\}^2 \qquad (9.9)$$

The choice of distribution for the exercise time is up to the user. Example possible distributions are uniform where exercising is equally likely at any time up to T, a declining triangular distribution where the value of the option is perceived as decreasing with time, or a rising triangular distribution where exercising is more likely the closer time gets to T. The case $p_t = 1$ and $p_t = 0$, t = 0, 1, 2, ..., T–1 is the European option.

9.7 EXAMPLE CALCULATION OF OPTION VALUE

Let the exercise price of the stock be $10.20 at year 2. Let the interest rate be 10% per annum.

Estimates of moments. Estimates are required of the expected value and variance of the value of the stock at year 2. Based on past performance of the stock, the investor estimates/forecasts that the optimistic, most likely and pessimistic values of the stock will be $13.86, $9.90 and $5.94 respectively. Using Chapter 5 expressions for calculating expected values and variances from optimistic, most likely and pessimistic estimates, the expected value and variance of the value of the stock at year 2 are $9.90 and 1.74 (^{2}) respectively.

Present worth. The deterministic present worth of the investment is −$0.25, implying that the investment is not worthwhile using such a measure. However the variability produces an upside.

Discounting the exercise price (negative) and stock value (positive) at year 2 to the present, while using Equations (9.6) and (9.7),

$$E[PW] = -\$0.25 \text{ (as in the deterministic case)}$$

$$Var[PW] = 1.19 \text{ (\$}^2)$$

Φ **and OV.** Calculate the feasibility Φ, the probability that the present worth is positive. Calculate the mean of the present worth upside. (See Chapter 5.) Here $\Phi = 0.41$, and upside mean = $0.78, giving, from Equation (9.4),

$$OV = \$0.33$$

This is an estimate of the option value.

The above calculations are based on a normal distribution assumption for present worth and discrete time discounting. However any other distribution considered a suitable model for present worth can be used, and continuous time discounting can be used if preferred by the investor.

Comparison. For comparison, Black–Scholes gives an option value of $0.36. This is based on a volatility of 9.4%, calculated from Equation (A9.4), and S_0 obtained by discounting $9.90 at 10% per annum over 2 years. As a proportion of the exercise price of $10.20, the difference is 0.29%.

9.8 EXAMPLE IMPLEMENTATION OF THE APPROACH GIVEN

The following example further demonstrates the straightforward nature of the approach given. An example European call option is used.

Consider an exercise price of K = $13.00 at T = 12 months, and assume an interest rate of 10% per annum.

Estimates of moments. Estimates are required of the expected value and variance of the stock at T. Estimates might combine historical data, analysis, judgement and experience to come up with what are believed to be best estimates. More particularly, some example ways that such estimates might come about are given in Chapter 5. Alternatively, a proxy approach can be used if the investor is aware of similar stocks; or if the stock volatility has been estimated, then the variance can be established from Equation (A9.4), while in a risk-neutral world, the expected value of the future stock price might be estimated according to $E[S_T] = S_0 e^{rT}$ (Hull, 2006).

For this example, estimates at T of an expected stock price of $11.05 with a standard deviation of $2.81 (variance 7.88 ^{2}) are assumed.

Present worth. The stock price and exercise price are discounted to time 0. Using Equations (9.6) and (9.7), E[PW] = –$1.76 and Var[PW] = 6.45 (^{2}) (standard deviation of $2.54).

Φ and OV. Feasibility, Φ, the probability that the present worth is positive, can now be calculated. Examining the positive portion of the present worth distribution, its area, Φ = 0.244, and its mean is $1.49. Such values may be calculated using the equation for a normal distribution (as in this example) or for whatever distribution is assumed, or numerically approximated by dividing the area into many vertical rectangles or strips and calculating the area as the sum of the rectangles, and the mean from the moment of these rectangles (Chapter 5).

The option value, OV, follows from Equation (9.4), and is OV = $0.37. Black–Scholes gives C = $0.42 (based on a volatility of 0.25 derived using Equation (A9.4), and S_0 obtained by discounting $11.05 at 10% per annum over 12 months), a difference of $0.05 or 0.4% as a proportion of the exercise price. OV is an estimate of the call value.

The above calculations are based on a normal distribution assumption for present worth, and discrete time discounting. However any other distribution considered a suitable model for present worth can be used, and continuous time discounting can be used if preferred by the investor.

Implied variance. For a given option value and estimated expected stock price at T, the calculation can be turned around to calculate an implied variance (or standard deviation).

The Greeks. The Greeks, being sensitivity measures, are most easily obtained by varying the analysis inputs one at a time, and observing the corresponding change in the option value.

Numerical studies show reasonable consistency for each of the Greeks between Black–Scholes and the approach given. The consistency is better for call options rather than put options, and this can be explained by the difference in the probability distribution assumptions. Using a nonsymmetric distribution for present worth and preventing stock prices from going negative would improve the comparison.

However it is suggested that rather than using the esoteric Greeks, thinking more in terms of conventional sensitivity analysis would be more understandable in plain English. Here the sensitivity of the option value to small changes in the analysis inputs, namely K, $E[S_T]$, $Var[S_T]$, r and T, is examined. Each of these inputs, one at a time, is varied plus/minus a small amount and the resulting option value calculated. Then for a small change in an analysis input (Carmichael, 2013):

- If the change in option value is small, the option value is insensitive to that analysis input.
- If the change in option value is large, the option value is sensitive to that analysis input.

Bounds. For K = 0: Then, $\Phi \cong 1$; Mean of PW upside $\cong S_0$; call OV $\cong$ $1 \times S_0 = S_0$; put OV = 0.

For K large: Then, $\Phi \cong 0$; call OV $\cong 0 \times$ Mean of PW upside = 0; put OV = S_0.

9.9 PUT–CALL PARITY

Usual put–call parity arguments for a European option give (for example, Hull, 2006),

$$C - P = S_0 - Ke^{-rT} \tag{9.10}$$

with $E[S_T] = S_0 e^{rT}$, where the notation is as for Equation (A9.1), and P is the value of the put option. Black–Scholes satisfies Equation (9.10).

In the notation of the book, and using the subscript C to denote call, and the subscript P to denote put, Equation (9.10) becomes,

$$OV_C - OV_P = E[PW]_C \text{ or } -E[PW]_P \tag{9.11}$$

where PW is the present worth of all cash flows (S_T and K) at time 0.

Consider the example in Section 9.8. For the call option,

$$\Phi_C = 0.244$$

$$\text{Mean of upside}_C = \$1.49$$

$$OV_C = \$0.37$$

$$E[PW]_C = -\$1.76$$

For the put option, doing similar calculations, but reversing the signs of the cash flows,

$\Phi_P = 0.756$

Mean of upside$_P$ = \$2.81

$OV_P = \$2.13$

$E[PW]_P = \$1.76$

Substituting these values in Equation (9.11), the left-hand side equals $0.37 - 2.13 = -1.76$, while both right-hand sides equal -1.76. The put–call parity Equation (9.11) holds.

Equation (9.11) can be shown to hold generally and to be independent of any assumption on the present worth distribution. Using the symbol W to stand for PW, and $f_W(w)$ the associated probability density function,

$$OV_C - OV_P = \int_0^\infty wf_W(w)dw + \int_{-\infty}^0 wf_W(w)dw = \int_{-\infty}^\infty wf_W(w)dw$$

$$= E[PW]_C \text{ or } - E[PW]_P \qquad (9.12)$$

The put–call parity Equation (9.11) holds.

9.10 EXOTIC OPTIONS

The approach given in this book is readily extendable to nonvanilla options when the scenarios are reduced to a collection of cash flows characterized by their expected values and variances. Examples of exotic options include those named as barrier, floating-strike lookback and exchange, to name a few.

Exotic options (exotics) have reasonable popularity because they are highly customizable financial instruments. However, due to the unique pay-off conditions of each exotic option, options valuing has a highly dispersed subject matter, with a range of methods available for valuing each exotic option. Exotics that can be modelled analytically are commonly valued using complicated extensions of Black–Scholes, which in itself is criticized for its complexity and its 'black box' approach for valuing options. The approach given in this book, on the other hand, offers a unifying approach that allows an investor to intuitively value any exotic option. The book's approach gives essentially the same option value as the existing preferred methods of analysis for each exotic type. It is observed that the deviation between the existing methods and the book's approach generally increases with increasing values of the analysis inputs. Nonetheless, the book's approach appears to be a useful tool for investors to approximate the option value due to its relative simplicity, flexibility and unifying approach to valuing all exotic options.

Exotic options are modified versions of standard vanilla options, where investors have customized the cash flow structure or payoff conditions of a standard option investment in order to meet some specific hedging need or to allow the investor to express a more specific prediction of the future. An exotic option may involve multiple underlying assets, provide protection from foreign-exchange rates, allow increased flexibility by allowing the investor to modify certain elements of the option prior to expiration, allow the investor to take on more risk by specifying the pattern that the asset price must follow in order for the investor to be entitled to a payoff or any combination of these features.

An exotic option differs from a standard vanilla option in terms of how and when the investor receives the payoff. Exotic options introduce particular features and conditions tailored to the payoff patterns desired by individual investors to administer more appropriate hedging strategies, or to simply adopt higher risk for higher return. While all exotic options share a set of similar principles, such as the establishment of an exercise price, each exotic option has a unique payoff structure that is generally more complex than that of a European vanilla option. One of the major challenges of options today is to provide an approach that can be universally applied to value any exotic option. It may be the lack of such an approach that has inhibited the growth of exotic options despite the flexibility of exotic options to cater to highly specific hedging requirements.

The literature gives extensions to Black–Scholes, allowing for closed form analytical results to many European exotics. A range of different methods exists for each exotic option. There is debate and controversy over the appropriateness of many of the methods and the validity of the option value that each method calculates in different scenarios for different investors. As well, these extensions are more complicated to use than Black–Scholes, and also rely on very sophisticated mathematical derivations, a background that most users do not have. Thus, investors have no intuitive understanding of the valuation process. Similarly, when valuing American exotics and exotics with complex payoff conditions, binomial lattices and numerical simulation of the underlying asset price movement over time provide nothing more than the final answer; the complexity of the process denies the user from intuitively understanding the value of the option. Furthermore, the inaccessible nature of these processes prevents the user from appropriately adjusting the calculated option value based on current information, such as current economic conditions and predictable future prospects of a business.

By contrast, the approach adopted in this book is very straightforward and provides a universal approach, while providing enough flexibility to allow the investor to intuitively adjust the value of the option by incorporating information on the current market/business climate.

The book's approach values an option by analyzing it as an investment with a series of cash inflows and outflows. Many exotic options have an identical cash flow structure to a standard vanilla option, that is, at time T (for a call option), a cash inflow of S_T and a cash outflow of K. Some have an additional cash flow in terms of a rebate, R. The payoff factors can relate, for example, to the path taken by one or more underlying asset values. These factors differ from one exotic option to another.

Example. A barrier option has a predefined barrier price, H, which the underlying asset price must breach at any time over the course of the option's life in order to entitle the investor to a payoff. If the barrier is breached, the option is treated as a standard vanilla option. If the barrier is not breached, the investor is not entitled to a payoff but may be entitled to a rebate, R, depending on the terms of the contract.

Now, OV = E*[PW], assuming that the investment will only go ahead if it is worthwhile (denoted *). The present worth is comprised of the asset price, exercise price, rebate and any other included cash flows. To apply this book's approach, each cash flow needs to be identified in terms of its value and timing, and how these are conditioned on other events occurring.

9.11 CARBON OPTIONS

With emission trading scheme (ETS) and carbon markets, carbon options have developed. Commonly, carbon options might be used in hedging against changing carbon prices, for example, for a company required to satisfy some carbon allowance, or in conjunction with Kyoto flexibility mechanisms such as the Clean Development Mechanism (CDM) where the viability of a project comes not only from the sale of an end-product but also from saleable carbon credits.

Commonly, the Black-76 method, a version of the Black–Scholes equation, might be used to value carbon options. The essential difference between Black–Scholes and Black-76 is that the latter uses a carbon futures price rather than the carbon current price. As such, the Black-76 method, where the futures price is equivalent to S_T, gives essentially the same option values as this book's approach.

The valuation of carbon options via the book's approach is no different from that given already for financial options. However the given approach offers a distinct advantage over existing methods in that the carbon price used in the calculations is not constrained to move by Brownian motion or any other model. The given approach allows the incorporation by the investor of current information, such as current economic and carbon market conditions and predictable future movements.

9.12 EMPLOYEE STOCK OPTIONS

Employee stock options may be evaluated as for other financial options by this book's given approach.

9.13 RANDOM CASH FLOWS AND INTEREST RATES

Chapter 10 gives present worth relationships for the cases where both cash flows and interest rates contain uncertainty. These may be used to find OV for this case.

9.14 ADJUSTMENTS

The book's approach is an alternative to Black–Scholes. Over the 40 years since the Black–Scholes equation was first published, users have come to understand its strengths and weaknesses, and its idiosyncrasies such as how it over- or undervalues calls out-of- or in-the-money, and they have learnt how it needs adjusting to deal with some of its failures. With the approach given, refinement will be required, for example, in the choice of present worth distribution, selection of stock price variance, selection of interest rate and opting for continuous or discrete time discounting versions. These refinements will be ongoing as users test it.

The approach given is ideally suited to real options but will need refinement for financial options.

Numerical testing has been carried out over a wide range of analysis inputs to establish confidence in the approach. However, as with any numerical testing, this represents spot testing and not universal testing. Further numerical testing could be examined. However, based on the significant amount of testing already carried out, the approach given is believed to be robust.

APPENDIX: RELATIONSHIP TO BLACK–SCHOLES

A9.1 Introduction

The approach given is suggested as an alternative to Black–Scholes and to be used to estimate the option value. In structural terms, the approach given is shown to perform the same calculation as Black–Scholes and capture the upside of an investment in the same way. Numerical testing, over a range of analysis inputs, confirms that the two approaches calculate essentially the same thing and result in essentially the same option values.

It is argued that the main advantages of the approach given over Black–Scholes are that it is straightforward to use, is intuitively appealing and requires a minimal background in mathematics. Users are able to visualize the upside of any investment, while the approach given represents no more than an extension to established and well-known deterministic present worth analysis. While implementation of Black–Scholes is straightforward as a black/grey box, understanding its derivation, assumptions and the manner in which it calculates an option value are not so straightforward.

A9.2 Background

Black–Scholes has been around for approximately 40 years, is well accepted and used and is a standard inclusion in university courses. Publications examining the performance, extensions, deficiencies and applications of Black–Scholes number in the thousands. Many adjustments and improvements to Black–Scholes have been published. Generally, these adjustments and improvements rely on the same basic approach as Black–Scholes; they are contrasted with the completely different line of attack of the approach given in this book. Binomial lattices and numerical simulation of the underlying asset price movement over time can also be used to value options, and are regarded as giving equivalent results to Black–Scholes.

Black–Scholes relies on a number of assumptions:

- Volatility is constant over time.
- Stock price movements follow geometric Brownian motion, and stock prices are lognormally distributed.
- The stock pays no dividends.
- The interest rate is constant and risk-free.
- The option can be exercised at expiry (European option).

Such assumptions may not necessarily represent accurately what is observed; however, Black–Scholes is still used in options trading as a comparison tool to establish whether an options contract is over- or undervalued. Both the risk-free rate and volatility could be anticipated to vary with market conditions. Gross movements in stock prices may be associated, for example, with company reorganizations. Dividends might be paid prior to expiration. Adjustments have been proposed to overcome many of the issues such as constant volatility, early exercising and incorporating dividends, but still Black–Scholes remains popular because of its simplicity of use.

The approach given here allows extensions beyond some of the restrictive assumptions behind Black–Scholes. Both Black–Scholes and the approach given value the upside of an investment similarly. Because exercising the option is a right and not an obligation, the exercising takes place only if it is worthwhile or has an upside to the investor. Downsides, implying that the investment is not worthwhile, do not come into the investor's calculations. The option holder is in effect protected against losses but rewarded for gains. Increased uncertainty in the stock price and time to expiration raise the upside of the investment (Hull, 2006).

The basis of the approach given in this book is no more than a conversion of any investment into a collection of cash flows characterized by their expected values and variances, and examining the present worth of these cash flows. The approach given permits whatever cash flows, and uncertainties attached to these cash flows and other analysis input variables, that the investor wishes to incorporate. Further comment is made on these characteristics below.

A9.3 Black–Scholes

Although relatively straightforward to apply, the derivation and understanding of the Black–Scholes equation require a high degree of mathematical sophistication, being based on Ito stochastic calculus and geometric Brownian motion.

Black–Scholes values a European call option according to (Hull, 2006),

$$C = N(d_1)S_0 - N(d_2)Ke^{-rT} \tag{A9.1}$$

where

$$d_1 = \frac{\ln\left(\frac{S_0}{K}\right) + (r + 0.5\sigma^2)T}{\sigma\sqrt{T}}$$

$$d_2 = d_1 - \sigma\sqrt{T}$$

and,

C   the value of the call option
$N(d_i)$ the cumulative standard normal distribution of the variable d_i
S_0 the stock price at time 0
r   the risk-free rate; Black–Scholes uses continuous time discounting
K   the price of exercising the option; strike price
T   the option life; the (expiry) time available to exercise the option; the time to maturity of the option
σ   stock price volatility

Put options can be similarly considered.

Nielsen (1992), among others, gives an explanation of the terms in Black–Scholes. In particular, the first term: $N(d_1)S_0$ is the expected (deterministic) present worth of the stock if the option finishes in the money, that is, if the stock price is above the exercise price; $N(d_1)$ is the risk-adjusted probability that the stock price will finish above the exercise price at the expiration date. And the second term: $N(d_2)Ke^{-rT}$ is the expected (deterministic) present worth of the exercise price if the option finishes in the money; $N(d_2)$ is the risk-adjusted probability that the option will be exercised; Ke^{-rT} is the present worth of the exercise price.

In effect, Black–Scholes calculates the discounted payoff that would occur if the current stock price increases above the exercise price, adjusted for the probability that the stock price will be greater than the exercise price. That is, Black–Scholes (Equation A9.1) effectively calculates the option value as,

$$C = \text{Expected* PW of stock price} - \text{Expected* PW of exercise price} \quad (A9.2)$$

assuming that the investment is only made if it is worthwhile (in the money); the symbol * is used to denote this assumption.

A9.4 Differences with Black–Scholes

C in Black–Scholes and OV in the approach given are obtained through expressions that are structurally the same, and OV becomes an estimate of C.

Apart from the large disparity in required mathematical sophistication between the approach given and Black–Scholes, there are differences in assumptions.

Distribution assumptions. Black–Scholes is based on a lognormal distribution assumption for stock prices. This is compared with the approach given, which makes no assumption on the distribution of the stock prices. Stock prices are described only in terms of expected values and variances. (For present worth, however, the user selects whatever distribution they think is most appropriate; some people might select a normal distribution.)

Incorporation of uncertainty. Both methods incorporate uncertainty or variability, but each does it in a different way. Black–Scholes incorporates uncertainty through volatility, while the approach given incorporates uncertainty within the stock price variance; the two can be related as mentioned below such that high volatility is associated with high stock price variance and vice versa. Increasing the volatility and increasing stock price variance both lead to an increase in the value of an option, and vice versa. Stock price variance could be anticipated to increase with time.

Discounting assumptions. Black–Scholes uses continuous time discounting. The formulation for the approach given is given above in terms of discrete time discounting. However a continuous time discounting version can be accommodated.

Current stock price, S_0. Unlike Black–Scholes, the approach given only uses the current stock price indirectly; instead the approach given works with the present worth of an estimate of the stock price at the time of exercising. However, the current stock price may play a role in the approach given in establishing an estimate of future stock prices, for example in a risk-neutral world (Hull, 2006), the expected value of the future stock price might be estimated according to $E[S_T] = S_0 e^{rT}$.

A9.5 Numerical comparison

All numerical example cases examined demonstrate that the approach given determines values that are essentially the same as Black–Scholes. Numerical studies show reasonable consistency between Black–Scholes and the approach given for the relationship between option value and: stock price, exercise price, time to exercising and volatility. Using a nonsymmetric distribution for present worth, continuous time discounting and preventing stock prices from going negative, would improve the comparison.

Indicative results of numerical studies on European call options are shown here. The option values calculated via the approach given and Black–Scholes are compared here over a range of analysis input variables:

Interest rates (% p.a.): $0.05 \le r \le 0.15$

Volatility (%): $0.10 \le \sigma \le 0.50$

Time to expiration (months): $1 \le T \le 24$

Interest rates, volatilities, times to expiration and exercise prices are changed in different combinations in the numerical testing, for investments close to and far from the money. Normal distributions for present worth are used as an example. Figures 9.3 to 9.6 show the differences between the values calculated by Black–Scholes and the approach given, that is $C - OV$, expressed as a percentage of the exercise price K. Normalization with respect to the exercise price is used in order to remove the influence of different magnitudes of stock prices.

The variance of the stock price at T is calculated from (Hull, 2006),

$$Var[S_T] = E[S_T]^2 \left(\exp(\sigma^2 T) - 1 \right) \tag{A9.3}$$

or

$$\sigma = \sqrt{\frac{\ln\left(\dfrac{Var[S_T]}{E[S_T]^2} + 1 \right)}{T}} \tag{A9.4}$$

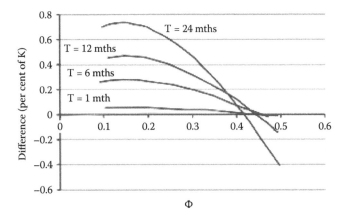

Figure 9.3 Difference (per cent of K) versus feasibility, Φ; σ = 0.25, r = 0.1.

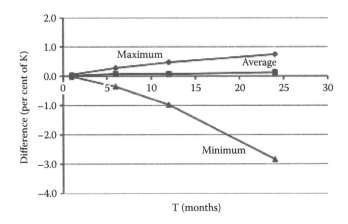

Figure 9.4 Difference (per cent of K) versus time to expiration, T.

Figure 9.3 shows the comparison as feasibility, Φ, changes, that is as the investment goes from far from the money to close to the money, for common volatility and interest rate assumptions. The difference is less than 0.8%. The approach given may be a better predictor than Black–Scholes in deep out-of-the-money calls and deep in-the-money calls.

Figures 9.4 to 9.6 show the comparison as time to expiration, T, volatility, σ and interest rate, r, change. Each of these figures gives envelopes – plots of maximum positive difference and maximum negative difference – and plots of average difference. Note that the differences in all the figures are not additive, but are the same differences represented against Φ and each analysis input variable in turn. Maximum differences

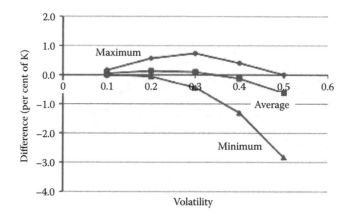

Figure 9.5 Difference (per cent of K) versus volatility, σ (per cent).

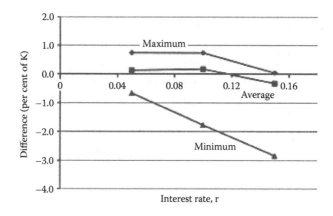

Figure 9.6 Difference (per cent of K) versus interest rate, r.

occur with large T, σ and r. For small T, σ and r, and averages, there is essentially no difference between the two methods.

All numerical testing demonstrates that the approach given and Black–Scholes calculate essentially the same value.

Exercises

9.1 With the book's approach, the distribution for S_T is replaced with $E[S_T]$ and $Var[S_T]$, that is in terms of moments only. While the shape of the distribution for S_T can be reflected in the choice of optimistic, pessimistic and most likely values, this knowledge of the shape is lost when using two moments only. A third moment of skewness is necessary to incorporate knowledge of the distribution shape. A distribution with a tail to the right has a positive skewness, with a tail to the left a negative skewness and a symmetrical distribution a zero skewness.

And so if a particular distribution is wanted for S_T, for example a lognormal distribution, this cannot be obliged. A flow on from this is that if S_T can only take positive values, then this can only be guaranteed through manual intervention, or where stock prices have expected value and variance magnitudes such that the probability of the stock price being negative is very small.

How might a lognormal shape for S_T be incorporated into the approach given?

Also consider: A lognormal distribution for the present worth of a stock price might be used. This would require separately from this (for a call option) the (deterministic) present worth of the exercise price to be subtracted. Is this workable? A lognormal distribution could not be fitted to $S_T - K$ collectively, because this could take negative values.

9.2 Is it possible to first evaluate OV_T and then discount this to time 0 to give OV, rather than discounting $E[S_T]$ and $Var[S_T]$ to time 0, and then evaluating OV? Would you anticipate that you would get the same answers either way that you did the calculation? Is OV the same as a discounted OV_T? If not, what is the difference and why is there a difference? Test this numerically. Try a range of values of S_T, K, r and T. Compare OV calculated by both approaches.

9.3 Using a nonsymmetric distribution for present worth, and preventing stock prices from going negative, would improve the comparison between Black–Scholes and the approach given. What distribution, alternative to a normal distribution might be suitable for present worth? How can stock prices be prevented from going negative?

9.4 The numerical testing of the approach given against Black–Scholes is for call options. Do similarly for put options. Would you expect similar comparison accuracy between call and put options? What influence does the assumption of a normal distribution have on the values calculated for a call option compared to a put option?

9.5 Do a numerical comparison of the behaviour of the Greeks for Black–Scholes compared with the approach given. What do you conclude?

9.6 What adjustments/refinements would you recommend in order that the book's approach gets closer to the results from Black–Scholes?

9.7 What adjustments/refinements would you recommend in order that the book's approach gets closer to the behaviour observed in the stock market?

9.8 Examine Black–Scholes and the book's approach when both are at the money. What do you see?

9.9 The approach given mentions nothing about arbitrage. Should it?

9.10 For the approach given, instead of talking of Greeks, which have no direct meaning within the approach given, what sensitivities would you recommend calculating instead? Think of sensitivities that people can intuitively understand.

Chapter 10

Probabilistic cash flows and interest rates

10.1 INTRODUCTION

This chapter explores the relationship between interest rate uncertainty and investment value. Generally, uncertainty in interest rates increases the value of an investment. This is demonstrated in a readily understandable way using a second order moment analysis. Two case studies – a long-term infrastructure project, and short/mid-term financial options – are given as examples. A method for showing equivalence between fixed and variable rate loans is demonstrated.

Researchers have attempted to incorporate interest rate uncertainty through reasonably complicated mathematical models of time-varying interest rates, and mathematical analysis that, unfortunately, is not understood by most practitioners, who prefer in the alternative to use simplifying assumptions and deterministic analysis. Yet, particularly for long-life investments, this latter way is questionable. Usual deterministic analysis undertaken by practitioners deals with uncertainty separate from the base investment analysis, via sensitivity analysis, scenario analysis or using artificial risk-adjusted discount rates. Sensitivity analysis is unable to differentiate grades of separation from a base investment value, while a scenario analysis is less definite again. Risk-adjusted discount rates attempt to compensate for a range of uncertainties in the analysis input variables, but particularly in the cash flows; the debate on the applicability of risk-adjusted discount rates is well documented. Deterministic analysis provides only one realization, where many could be anticipated because of the uncertainty in rates.

The impact of interest rates on investment analysis has been looked at widely in the literature; it is acknowledged that the value of an investment depends on the variability of interest rates over time. What is lacking heretofore is a straightforward result on the effects of interest rate uncertainty on both long- and short-term investments.

With options, the Black–Scholes equation assumes a constant interest rate. Short-term investments may not be much affected by such an assumption.

However, for longer-term investments, the presence of interest rate volatility may lead to discrepancies between estimate and actual. Generalizations of Black–Scholes equation to include an interest rate modelled by a stochastic differential equation exist. Uncertainty in interest rates increases the option value. This is demonstrated below in an alternative and readily understandable way using a second order moment analysis.

10.2 INVESTMENT ANALYSIS

The analysis of this chapter is an extension of Chapter 6, allowing interest rates to be random variables. The uncertainty in the interest rate is incorporated through its variance.

The formulation covers both conventional investment, and options (real and financial), and can be specialized to suit particular situations. It is written in terms of assets, where assets may be financial or real. In simple terms, the basis of the formulation is a conversion of any investment into a collection of cash flows characterized by their expected values and variances, and an interest rate characterized by its expected value and variance. Where the variances are zero, the following reduces to a conventional deterministic present worth treatment.

10.2.1 General investments

Consider a general investment, with possible cash flows extending over the life, n, of an investment. The net cash flow, X_i, at each time period, i = 0, 1, 2,..., n, may be the result of a number of cash flow components (random variables).

The present worth, PW, is

$$PW = \sum_{i=0}^{n} PW_i = \sum_{i=0}^{n} \left[\frac{X_i}{(1+r)^i} \right] \tag{10.1}$$

where r is the interest rate, and PW_i is the present worth due to X_i, i = 0, 1, 2, ..., n. And,

$$E[PW] = \sum_{i=0}^{n} E[PW_i]$$

$$Var[PW] = \sum_{i=0}^{n} Var[PW_i] + 2\sum_{i=0}^{n-1}\sum_{j=i+1}^{n} Cov[PW_i, PW_j] \tag{10.2}$$

Alternatively, the variance expression can be written in terms of the intertemporal correlation coefficients, ρ_{ij}, between PW_i and PW_j,

$$\text{Var}[PW] = \sum_{i=0}^{n} \text{Var}[PW_i] + 2\sum_{i=0}^{n-1}\sum_{j=i+1}^{n} \rho_{ij}\sqrt{\text{Var}[PW_i]}\sqrt{\text{Var}[PW_j]} \qquad (10.3)$$

For independent PW_i,

$$\text{Var}[PW] = \sum_{i=0}^{n} \text{Var}[PW_i] \qquad (10.4)$$

For perfect correlation of the PW_i,

$$\text{Var}[PW] = \left(\sum_{i=0}^{n} \sqrt{\text{Var}[PW_i]}\right)^2 \qquad (10.5)$$

The contributions that $E[PW_i]$, $\text{Var}[PW_i]$ and $\text{Cov}[PW_i, PW_j]$ make to these expressions can be developed further using first order and second order approximations based on a Taylor series (Chapter 5).

Given that,

$$PW_i = \frac{X_i}{(1+r)^i} \qquad (10.6)$$

and using a second order approximation based on a Taylor series (Chapter 5), where X_i and r are uncorrelated random variables,

$$E[PW_i] = \frac{E[X_i]}{(1+E[r])^i} + \frac{i(i+1)E[X_i]}{2(1+E[r])^{i+2}}\text{Var}[r] \qquad (10.7)$$

Equation (10.7) gives directly the relationship between the value of an investment and interest rate variance. For a given net positive cash flow, Equation (10.7) indicates that an increase in the variance of the interest rate increases the expected value of PW_i, which in turn increases the expected value of PW, and hence the feasibility Φ. For a given net negative cash flow, the situation reverses. In terms of risk attitudes, people usually want a higher return from an investment with higher uncertainty, but counter to this, some people may be less likely to invest as the degree of uncertainty increases. In terms of the current practice of increasing hurdle rates to take care of uncertainty, Equation (10.7) implies the opposite, namely that hurdle rates should be decreased with increased uncertainty.

Summary result 1a: For general investments, for a given net positive cash flow, an increase in the variance of the interest rate increases the present worth of the investment. This influence of interest rate variance is increased with increasing life of the investment and increasing size of expected cash flows. More investments could be anticipated to become viable with increasing uncertainty in the interest rate. For a given net negative cash flow, the situation reverses.

Using a first order approximation based on a Taylor series, for X_i and r uncorrelated,

$$\text{Var}[PW_i] = \frac{1}{\left(1+E[r]\right)^{2i}} \text{Var}[X_i] + \left(\frac{i^2 E^2 [X_i]}{\left(1+E[r]\right)^{2i+2}} \right) \text{Var}[r] \qquad (10.8)$$

Equation (10.8) gives directly the relationship between the present worth variance and interest rate variance. For a given net cash flow (positive or negative), Equation (10.8) indicates that an increase in the interest rate variance increases the PW_i variance, which in turn increases the PW variance. Intuitively this result is sound, because it is reasonable that an increase in the variance of an analysis input random variable (r) would increase the variance of the overall function (PW).

Summary result 1b: *For general investments, for given cash flows, an increase in the variance of the interest rate increases present worth variance. This influence of interest rate variance is increased with increasing life of the investment and increasing size of expected cash flows.*

10.2.2 Financial options

The above approach incorporates the various option cases. For example, consider a European call option, where S_T is the asset value (stock price), and K is the exercise price at time T, then the only cash flows are at i = T. That is,

$$X_T = S_T - K \qquad (10.9)$$

A put option has the signs reversed. The exercise price K is taken as being deterministic, that is its variance is zero. S_T is a random variable. S_T and K are independent. The present worth (at time 0), PW, is only due to X_T and is

$$PW = \frac{S_T - K}{\left(1+r\right)^T} \qquad (10.10)$$

Using Equation (10.7),

$$E[PW] = E[PW_T] = \frac{E[S_T] - K}{\left(1+E[r]\right)^T} + \frac{T(T+1)\left(E[S_T] - K\right)}{2\left(1+E[r]\right)^{T+2}} \text{Var}[r] \qquad (10.11)$$

Using Equation (10.8),

$$\text{Var}[PW] = \text{Var}[PW_T] = \frac{1}{\left(1+E[r]\right)^{2T}} \text{Var}[S_T] - \left(\frac{T^2\left(E[S_T]-K\right)^2}{\left(1+E[r]\right)^{2T+2}}\right)\text{Var}[r]$$

(10.12)

The feasibility is the probability that the present worth is positive, $\Phi = P[PW > 0]$. A Φ value close to 0 is considered as being far out of the money, whereas a value close to 1 is considered as being far in the money. Having characterized the present worth using an appropriate distribution, then the option value estimate is given by the Carmichael equation (Chapter 7),

$$OV = \Phi \times \text{Mean of PW upside}$$

Equations (10.11), (10.12) and the Carmichael equation give directly the relationship between the option value and interest rate variance. Assuming that the option is in the money at expiry ($S_T - K > 0$), then Equation (10.11) indicates that an increase in interest rate variance increases the expected value of PW and hence increases Φ, which in turn increases the option value. Equation (10.12) indicates that an increase in interest rate variance increases PW variance, and hence increases the 'Mean of PW upside', which in turn increases the option value.

Summary result 2. Where the expected stock price at expiry is greater than the exercise price, the value of a call option increases with increasing interest rate variance. This influence of interest rate variance is increased with increasing difference between the expected stock price and the exercise price, and increasing time to expiry. Other option types are anticipated to behave similarly.

10.2.3 Real options

The above approach can be adapted to the various real option cases. For example, an option to expand is viewed as a call option; an amount is spent now in order to have flexibility in the future – a future sum (exercise value, K) would be paid to expand an asset/facility and provide an ongoing (beyond T) return. At time T, the equivalent of the stock price S_T is the present worth (at time T) of the subsequent cash flows resulting from exercising the option.

Summary result 3. Where the expected present worth at T of future cash flows is greater than the exercise amount, the value of an option to expand increases with increasing interest rate variance. This influence of interest rate variance is increased with increasing difference between the expected present worth at T of future cash flows and the exercise amount, and increasing time to expiry. Other option types are anticipated to behave similarly.

10.3 CASE STUDY – GAS TRANSMISSION PIPELINE

A gas transmission pipeline is used here as an example infrastructure project. Such infrastructure is characterized by significant capital investment, long life and tends to be single purpose in nature.

Gas transmission pipelines transport natural gas from production fields to major demand centres such as a city, town, power generation facility or a mine. Transmission pipelines are characterized by wide diameters and operate under high pressure. A long-term arrangement for gas supply is contracted prior to any investment in constructing the pipeline. This long-term security on supply gives greater certainty in the forecast usage levels and enables reasonable estimation of cash flows over the life of the project.

Pipeline revenue is derived from the tariff charged for the transportation of the gas. A pipeline's tariff reflects the transportation distance, capital costs of the pipeline and associated facilities, the pipeline's age and extent of depreciation, the geography along the pipeline route and the availability of spare capacity in the pipeline.

The capital cost represents a large cash outflow at the beginning of the project life. Its components include costs associated with: the pipeline, compressor stations, right of way (ROW) acquisitions, mainline valve stations, meter stations, pressure regulator stations, spares, environmental, SCADA (supervisory control and data acquisition), telecommunications, engineering, procurement, and construction management (EPCM) and development costs (planning and other approvals, planning and environmental compliance costs, legal costs, bank fees, owner-arranged insurances, specialist design and technical studies). For the approximately 500 km pipeline with a lifespan estimate of 50 years used in this study, the total project (deterministic) cost estimate is $560M. Pessimistic, most likely and optimistic estimates for the capital cost components are estimated.

A draw-down schedule is used to determine the financing costs for the initial three-year period, of which the second and third years are construction. Straight line depreciation is used over component lifetimes. A zero salvage value at the end of component lifespans is assumed, except for two compressor stations and their spares. Operating expenditure (OPEX) is assumed to grow with time. Maintenance capital expenditure is assumed as a percentage of total revenue initially, then reduces in line with depreciation. Additional compressor stations are to be added during the life of the project. The reservation capacity is assumed to grow slightly each year to meet forecast demand for gas transmission. The reservation revenue has zero variance based on defined contracted amounts. Average annual throughput is taken as 90% of the annual reservation capacity.

The (base case) interest rate expected value (6.27% per annum) and standard deviation (1.37% per annum) are obtained from historical data, with data from the peak of the global financial crisis (GFC) excluded. The corporate tax rate is 30%, and inflation 2.5% per annum.

FCFF (free cash flows to firm) are used for the X_i and comprise components. In determining $Var[X_i]$, the intercomponent correlation coefficients need estimating. PW comprises components PW_i. In determining $Var[PW]$, intertemporal correlation coefficients need estimating.

10.3.1 Influence of interest rate uncertainty

Compared with the deterministic interest rate case (standard deviation 0% per annum), the base probabilistic interest rate case outlined above (standard deviation 1.37% per annum) increases the expected present worth by 13.9%. From an investor's perspective, this result is important. Its importance is further emphasized in Figure 10.1.

Figure 10.1 shows the influence of $E[r]$ and $Var[r]$ on $E[PW]$. The results show that, as the interest rate variance increases, the expected present worth of the investment also increases. Figure 10.1 shows a decreasing slope for $E[PW]$ as $E[r]$ increases. The interest rate variance has a larger relative impact on $E[PW]$ at lower values of $E[r]$.

Figure 10.2 shows the influence of $E[r]$ and $Var[r]$ on $Var[PW]$. The results show that as the interest rate variance increases, the present worth variance increases. Similarly to the results for $E[PW]$, as $E[r]$ increases, the line

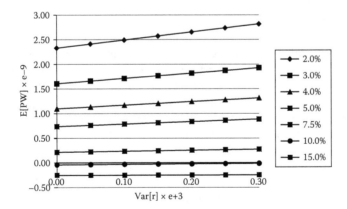

Figure 10.1 Influence of Var[r] on E[PW], for E[r] ranging from 2% to 15% p.a. (From Carmichael, D. G. and Bustamante, B. L., Interest Rate Uncertainty and Investment Value – A Second Order Moment Approach, School of Civil and Environmental Engineering, The University of New South Wales, Sydney, Australia, 2014.)

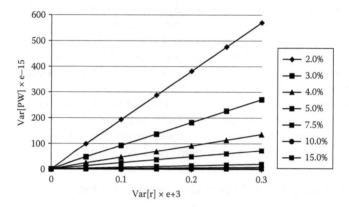

Figure 10.2 Influence of Var[r] on Var[PW], for E[r] ranging from 2% to 15% p.a. (From Carmichael, D. G. and Bustamante, B. L., Interest Rate Uncertainty and Investment Value – A Second Order Moment Approach, School of Civil and Environmental Engineering, The University of New South Wales, Sydney, Australia, 2014.)

slopes decrease. The interest rate variance has a larger relative impact on Var[PW] at lower values of E[r].

10.4 CASE STUDY – INDEX OPTION

A European index option is used here as an example of a short-term investment. The value of an index option varies according to movements in the value of an underlying index (for example, based on a collection of stocks from companies with large market capitalization) in the same way that the value of a share option varies with movements in the value of the underlying shares.

The historical price data are expressed in points where one point equates to $10. The data are used as a basis for determining reasonable estimates for the expected value and variance of the index price at expiry, S_T.

Historical data for the index options are sourced from price history reports. These reports provide detailed historical information on options including the option code, expiry date, strike price, last trading date, volume traded and underlying price.

Three- and six-month call options selected from the price history report are used in the analysis. Hypothetical options have also been valued with maturities of years, in order to examine the change in impact of Var[r] as the time to expiry of an option increases. Investments close to and far from the money are considered. The option values are calculated according to the Carmichael equation (together with a second order moment analysis) and compared with that obtained by the Black–Scholes equation; differences between that approach and Black–Scholes values are expressed below as a percentage of the exercise price.

Interest rates are based on historical rates for durations leading up to the option purchase date with adjusting for investor expectations for the life of the option; these durations are matched to the relevant option expiry period. Optimistic, most likely and pessimistic values for S_T are estimated by observing historical ranges in the index over similar time periods to those of the options studied, combined with making appropriate judgements through the perspective of an investor.

Sensitivity-style studies are given for a range of interest rate variances. This range is bounded by a minimum of zero and a maximum value estimated from particularly volatile periods historically such as the global financial crisis (GFC) and the late 1980s, and based on time periods matching that of the option times to expiry. Generally, the interest rate variance increases as the time period examined increases.

Three-month (90-day) option. Two option cases are considered: one far from the money (K = 4000 points) and one close to the money (K = 4200 points); everything is the same for the two cases, except for the exercise price. Using historical 90-day rates, estimates for E[r] and Var[r] are respectively 4.66% p.a. and $(0.09\% \text{ p.a.})^2$. For estimated $E[S_T]$ and $Var[S_T]$ values of 4179 and 108^2 respectively, and Var[r] = 0 (no uncertainty), then $\Phi = 0.95$, OV = \$1786 (−0.68% difference as a per centage of K, compared with Black–Scholes) for K = 4000; and $\Phi = 0.42$, OV = \$328 (−0.33% difference as a percentage of K, compared with Black–Scholes) for K = 4200.

Figure 10.3 shows the influence of Var[r] on the option value. The range of interest rate variances identified as appropriate for a 3-month investment is 0 to $(1.00)^2$. The two highest levels of variance shown represent levels reached during the abnormal economic periods of the GFC and

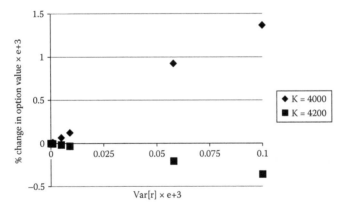

Figure 10.3 Three-month (90-day) option. Influence of Var[r] on option value; % change in option value relative to Var[r] = 0 case. (From Carmichael, D.G. and Bustamante, B.L., Interest Rate Uncertainty and Investment Value – A Second Order Moment Approach, School of Civil and Environmental Engineering, The University of New South Wales, Sydney, Australia, 2014.)

the late 1980s respectively. Figure 10.3 shows that the levels of interest rate variance have a very small influence on the value of the option.

Figure 10.3 shows a decreasing option value as the interest rate variance increases for the K = 4200 case. This is attributed to E[PW] becoming further negative as Var[r] increases, whereas in the K = 4000 case, E[PW] increases as Var[r] increases. The levels of interest rate variance for the close to the money case have a smaller influence on the value of the option, than in the far from the money case. On examination of Equation (10.12). it is seen that the influence of Var[r] is being diluted by the product term, namely $(E[S_T - K])$, which is smaller in the close to the money case. When $(E[S_T - K])$ is low, that is when close to the money, then the Var[r] has less of an influence on the option value than it does when far from the money.

Six-month (180-day) option. Consider two cases related to the 3-month cases just considered. Using historical 180-day rates, estimates for E[r] and Var[r] are respectively 4.96% p.a. and (0.19% p.a.)^2. For estimated E[S_T] and Var[S_T] values of 4267 and 167^2 respectively, and Var[r] = 0 (no uncertainty), then $\Phi = 0.94$, OV = $2631 (+0.32% difference as a percentage of K, compared with Black–Scholes) for K = 4000; and $\Phi = 0.65$, OV = $1018 (+0.21% difference as a percentage of K, compared with Black–Scholes) for K = 4200.

From Figure 10.4, as Var[r] increases, the option value increases. The higher anticipated levels of interest rate variance experienced over a 6-month investment period have a larger impact on the option value than those over a 3-month investment period. However the impact is still considered relatively small from an investor's perspective, for typical variance levels.

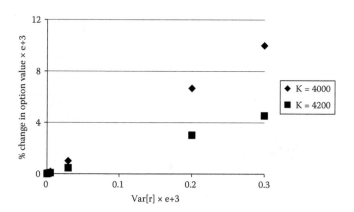

Figure 10.4 Six-month (180-day) option. Influence of Var[r] on option value; % change in option value relative to Var[r] = 0 case. (From Carmichael, D.G. and Bustamante, B.L., Interest Rate Uncertainty and Investment Value – A Second Order Moment Approach, School of Civil and Environmental Engineering, The University of New South Wales, Sydney, Australia, 2014.)

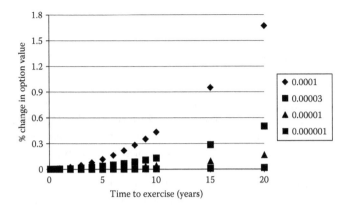

Figure 10.5 Influence of time to expiry on option value; % change in option value relative to Var[r] = 0 case, for Var[r] ranging from 0.000001 to 0.0001. (From Carmichael, D.G. and Bustamante, B.L., Interest Rate Uncertainty and Investment Value – A Second Order Moment Approach, School of Civil and Environmental Engineering, The University of New South Wales, Sydney, Australia, 2014.)

At lower Φ values, the increases in the option value are lower, because of lower values of the term $(E[S_T] - K)$.

Longer time options. Exchange traded options commonly have a maturity of less than or equal to 12 months. However, in order to further explore the influence Var[r] has on the option value as the time to expiry of an option increases, hypothetical options with longer times to expiry are examined. Similarly to the 6-month and 3-month options analyzed, when $E[S_T]$ is above the exercise price, increases in Var[r] increase the option value. The results show more impact of Var[r] on the option value than that seen in the options with shorter times to expiry, but still the influence is small. The size of the $(E[S_T] - K)$ term affects the magnitude of the impact on the option value.

Figure 10.5 shows the results for European call options, where $E[r] = 5\%$ (per annum,) K = 4500 points, $E[S_T]$ = 4767 points, and $\sqrt{Var[S_T]}$ = 333 points. The influence of Var[r] on the option value increases as the time to expiry increases. A low level of Var[r], typically as might be experienced over a 3-month expiry period, has a relatively negligible impact on the option value, irrespective of times to expiry. As Var[r] increases, it has a larger impact on the option value.

10.5 SUMMARY

Largely, the incorporation of interest rate variance directly into an investment analysis results in an increase in both expected present worth and variance of the present worth of the investment. Although it is found that the impact on short-term investments is very small, the impact on mid- and long-term investments indicates that interest rate variance can add value difference to an investment.

General investment. For general investments, for a given net positive cash flow, an increase in interest rate variance increases the present worth of the investment. This influence of interest rate variance is increased with increasing life of the investment and increasing size of expected cash flows.

Because infrastructure is characterized by significant capital investment and lifespans measured in decades, interest rate variance has a substantial influence on the investment value. Consistent with that for financial options, interest rate variance has a larger impact on the present worth of cash flows from time periods further into the future. A low expected interest rate coupled with high interest rate variance result in the most pronounced impact of interest rate variance.

The results of the gas transmission pipeline investment analysis support the view that interest rate variance should be directly incorporated into investment analysis. The influence of cash flows beyond 20 years is increased markedly by the uncertainty in the interest rate. Interest rate expected value and variance are key drivers of expected value and variance of the present worth of an infrastructure investment.

Financial options. Where the expected stock price at expiry is greater than the exercise price, the value of a call option increases with increasing interest rate variance. This influence of interest rate variance is increased with increasing difference between the expected stock price and the exercise price, and increasing time to expiry.

At low expected interest rate, the influence of interest rate variance is more marked. For short-term options, interest rate variance has only a minor influence on the value of the option. There is a moderate influence for mid-term options. Small increases in an option's value can be translated into considerable dollar amounts, where a large volume of options occurs.

Real options. Where the expected present worth at T of future cash flows is greater than the exercise amount, the value of an option to expand increases with increasing interest rate variance. This influence of interest rate variance is increased with increasing difference between the expected present worth at T of future cash flows and the exercise amount, and increasing time to expiry.

The above examples use a consistent expected interest rate and variance for the entire life of the investment. However, the investment could potentially be decomposed into stages. If, for example, an investor believes that interest rate variance is going to be particularly high for the first few years of the investment and subsequently revert back to a lower level for the remaining life of the investment, then this can be readily incorporated.

10.6 VARIABLE AND FIXED RATE LOAN EQUIVALENCE

A common situation is where a borrower is offered the choice of two types of loans – a fixed-interest rate loan for a fixed period, or a variable-interest (floating-interest) rate loan. Usual 'wisdom' and advice is that the borrower opts for the fixed-rate loan if interest rates are anticipated to rise in the future, or opts for the variable-rate loan if interest rates are anticipated to fall. However, depending on the magnitude of the rates offered, and the rate uncertainty in the market, such wisdom may not be well-founded. Here it is demonstrated why this might be. A readily usable, low-mathematical-background expression is derived showing when fixed- and variable-rate loans are equivalent, and when one rate type is better/worse than the other.

10.6.1 Introduction

When borrowing money, the loan may be structured using a fixed-rate and/or a variable- (floating-) interest rate. The magnitude and type of rate has a marked effect on the overall amount of interest to be repaid over the life of the loan. Commercial lenders have methods to determine a relationship between fixed- and variable-interest rate loans offered; however, these methods are not made public or made available to borrowers. And a search of the literature provides no help for borrowers. Accordingly, it would be helpful to provide borrowers with a straightforward way of establishing which is the better loan type to select in any given situation, or to know how much better/worse one rate type is than the other. This is the basis of what follows.

Market interest rates fluctuate over time and this fluctuation introduces uncertainty into any analysis. A fixed-rate loan means that the interest rate stays constant over the loan period, and hence the amount of interest to be repaid is unaffected by these fluctuations. This provides certainty in terms of loan repayments. On the other hand, a variable-rate loan has an interest rate linked to the fluctuations. If the market rate rises, the amount to be repaid will rise, while if the market rate falls, the amount to be repaid will be less (assuming that the lender passes on any interest rate movements).

To deal with fluctuating interest rates, variable-rate loans appeared in the 1970s. Prior to the introduction of variable-rate loans, investors and lenders may have used the strategy of rolling over short-term loans frequently, in order to readjust the interest rate. A variable-rate loan might therefore prove attractive to a borrower as well as a lender because it enables a single transaction to replace many separate transactions, thus reducing fees and administration costs.

The choice of loan type is particularly relevant for large capital expenditure projects, which generally require some debt financing, because interest payments will be large and these affect project profitability. But it applies equally at the other end of the investment scale, for example to homebuyers. Usual wisdom and advice is that the borrower opts for the fixed-rate loan if interest rates are anticipated to rise in the future, or opts for the variable-rate loan if interest rates are anticipated to fall. However, such wisdom, considering only trends and ignoring fluctuations and magnitudes, may not be well-founded. It also does not take into account the real possibility that the lender has already factored these trends or movements into the offered interest rates.

Having a straightforward fixed-variable rates equivalence relationship would therefore appear desirable. An expression showing when fixed- and variable-rate loans are equivalent is given below, and this provides the ability to establish when one offered rate type is better/worse than the other.

The development below is not about risk, or establishing equivalent risk between fixed- and variable-rate loans. To establish equivalent risk would require knowledge of the risk characteristics of the borrower and/or lender. Both variable-rate loans and fixed-rate loans have associated risk, derived from uncertainty in the future interest rate market. Variable-interest-rate loans have downside risk resulting from the possibility of market interest rates moving above quoted fixed rates, while fixed-interest-rate loans have downside risk resulting from the possibility of market interest rates moving downwards.

The development below is written in terms of loans for definiteness, but can be generally applied to any investment giving a fixed- or variable-interest return. It addresses one issue – the fundamental relationship between the fixed- and variable-interest rates, independent of other loan dynamics and features, and applicable to both lender and borrower.

10.6.2 Fundamental relationship

A second order moment analysis, whereby random variables are characterized by their expected values and variances, is used here because of the insight it gives. Uncertainty is incorporated through the variances.

Consider borrowing an amount Z now, where there is a choice of a fixed-rate loan or a variable-rate loan. With interest rate r, let Z grow to X_i at time i; i = 0, 1, 2, ... That is,

$$X_i = Z(1+r)^i \tag{10.13}$$

Using a Taylor series expansion, in conjunction with expectation, and keeping only up to second moments, then,

$$E[X_i] = Z(1+E[r])^i + \frac{1}{2}\left[Z(i)(i-1)(1+E[r])^{i-2}\right]Var[r] \tag{10.14}$$

$$Var[X_i] = Var[r]\left[Z(i)(1-E[r])^{i-1}\right]^2 \tag{10.15}$$

A Taylor series approximation is acceptable provided the coefficients of variation are not large and the function approximated is not too nonlinear (Benjamin and Cornell, 1970; Ang and Tang, 1975). See Chapter 5. Both provisions are satisfied here. Equations (10.14) and (10.15) show that the expected total cost of the loan and its variance increase with interest rate uncertainty or variance.

For the expected total cost of the loan to be the same (at time i), under both fixed and variable rates, then set Equation (10.13) (deterministic rate) equal to Equation (10.14) (stochastic rate) to give,

$$Z\left(1+F^{i}\right)=Z(1+V)^{i}+\frac{1}{2}\left[Z(i)(i-1)(1+V)^{i-2}\right]Var\left[r\right] \tag{10.16}$$

where r_F and r_V are the fixed and variable interest rates respectively over the loan period, $E[r_F] = F$ and $E[r_V] = V$. Equation (10.16) ignores any interim loan repayments that might be made, or assumes that interim repayments are deterministic and the same for each loan type.

Simplifying gives the required relationship between the rate types,

$$F=\sqrt[i]{(1+V)^{i}+\frac{1}{2}\left[(i)(i-1)(1+V)^{i-2}\right]Var\left[r_{v}\right]}-1 \tag{10.17}$$

10.6.3 Sensitivity analysis

The behaviour of Equation (10.17) can be illustrated with some sensitivity-style studies. Here the interest rate variance and loan period are altered and the difference between the fixed and expected variable rates examined.

Changing the rate variance. Figure 10.6 shows the influence of changing the estimated rate variance, for a given loan scenario. Similar trends occur for other scenarios.

The theoretically equivalent fixed rate is always greater than the expected variable interest rate. An increase in the variance of the interest rate results in an increase in the difference between the fixed-interest rate (F), and the expected variable-interest rate (V). As the variance of the interest rate increases, the gradients of the plots in Figure 10.6 become larger negative values; that is, for higher-rate variances, the difference between the fixed rate and the expected variable rate decreases more rapidly with increases in the expected variable rate.

The future worth of a present sum with uncertain interest rates is greater than that with deterministic rates. In order to compensate for this higher value, for equivalence, the fixed interest rate needs to be the higher. This is demonstrated in Figure 10.7.

As the variance of the interest rate increases, the expected future value also increases, and the difference between the fixed (rate variance of zero)

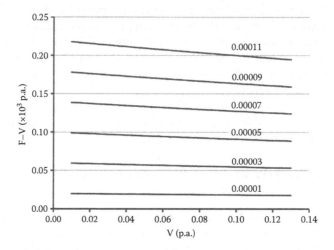

Figure 10.6 Example influence of changes in Var [r_v] = 0.00001 – 0.00011; i = 5 years. (From Carmichael, D. G. and Handford, L. B., Equivalent Fixed-Rate and Variable-Rate Loans, School of Civil and Environmental Engineering, The University of New South Wales, Sydney, Australia, 2014.)

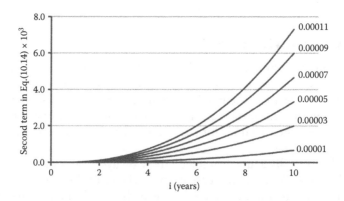

Figure 10.7 Example contribution of the second term in Equation (10.14); Z = 1, F = V = 5% p.a., Var [r_v] = 0.00001 – 0.00011 (% p.a.)^2. (From Carmichael, D. G. and Handford, L. B., Equivalent Fixed-Rate and Variable-Rate Loans, School of Civil and Environmental Engineering, The University of New South Wales, Sydney, Australia, 2014.)

and expected variable-interest rates increases. Thus, if the expected future values are to be equal, the fixed-interest rate must be increased in line with the variable rate variance.

Changing loan length. Figure 10.8 shows the influence of changing the loan length, for a given loan scenario. Similar trends occur for other scenarios.

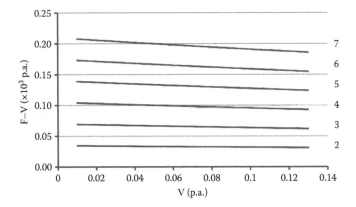

Figure 10.8 Example influence of changes in i = 2 – 7 years; Var[r$_v$] = 0.00007. (From Carmichael, D. G. and Handford, L. B., Equivalent Fixed-Rate and Variable-Rate Loans, School of Civil and Environmental Engineering, The University of New South Wales, Sydney, Australia, 2014.)

A time unit of year is used here but other time units, such as month, would need to be used for equivalence values for time less than one year, because for i = 1, F and V are the same using Equation (10.17).

Figure 10.8 shows that as i increases, the difference between the equivalent rates increases; that is, the longer the loan period, the greater the fixed rate should be. As the loan period gets larger, the gradients of the plots in Figure 10.8 become larger negative values; that is, the difference between the fixed and expected variable rates decreases more rapidly with increases in the expected variable rate.

10.6.4 Empirical support

Although extensive data are available from central and private banks on historical interest rates for business, term, home and other loans, direct empirical support for this book's result, namely Equation (10.17), is difficult to obtain because of the characteristics of available data sets:

- Averaging of rates across multiple loans occurs.
- Rates include bank risk margins; risk margins vary from borrower to borrower, lender to lender and loan to loan; the risk characteristics of borrowers and lenders are unknown.
- Loans may have different levels and types of security.
- The lengths of many loans are not published.
- The uncertainty in the interest rates at the time of a loan is not published; anticipated trends or movements in the interest rates at the time of a loan are not published.

- The rate offered may be different at the start of a loan to later in a loan.
- Different loans have different contractual features (termination, early repayment, offset, etc.); each loan can be tailored to the borrower and the circumstances.

It could be anticipated that lenders would determine their rates according to the competing lending and borrower markets, money availability and business reasons, and so raw publicly available data would not be available to support the book's theory.

What can be concluded from the available data sets, in support of this book's result, is that the relationship between the expected rates is essentially linear, and that an increase in the length of the loan (where data are available) results in an increase in the equivalent fixed-interest rate.

10.6.5 Example usage

Assume that a borrower is considering a loan over a period of 5 years. The borrower can choose either a fixed-rate loan or a variable-rate loan over this period.

The borrower estimates that over this period (based on historical interest rate movements, the anticipated economy, inflation, etc.), the most likely variable interest rate per annum applying to the loan would be 5% per annum, with optimistic and pessimistic values of 2.5% per annum and 6.5% per annum. From these, estimates of the interest rate expected value and variance can be obtained. Alternatively, the mean and standard deviation might be estimated directly, or one of a number of other techniques could be used to obtain the same information. The borrower would like to know what fixed-interest rate would be equivalent.

From the optimistic, most likely and pessimistic estimates (Chapter 5), $E[r_v] = V = (0.025 + 4 \times 0.05 + 0.065)/6 = 0.0483\%$ p.a., and $Var[r_v] = ((0.065 - 0.025)/6)^2 = (0.0067\%$ p.a.$)^2$. Substituting these values in Equation (10.17), gives $F = 4.84\%$ p.a. That is, if offered a fixed-rate loan with a lower rate than this, and all else being equal, this would be a more desirable loan than the variable-rate loan, and vice versa.

10.6.6 Summary

On the equivalence of fixed-rate and variable-rate loans, the following are found generally for all interest rates, rate variances and loan lengths:

- The fixed-interest rate should always be greater than the expected variable-interest rate.
- With increased rate variance and length of the loan, the difference between the fixed-interest rate and the expected variable rate increases, and the gradient of the difference in the expected rates increases.

Anticipated trends in interest rates can be accommodated in Equation (10.17) through estimates of expected values and variances in future interest rates. Popular 'wisdom' is that if interest rates are anticipated to rise, then a fixed-rate loan is preferable, and vice versa. Equation (10.17), however, shows that this may not necessarily be the correct way to think. Lenders presumably would be aware of this popular wisdom and would almost certainly factor this into any rates offered. A better strategy for borrowers would be to follow Equation (10.17) and the conclusions flowing from it. No one knows exactly how interest rates will move, but this uncertainty is embodied unemotionally in Equation (10.17).

10.6.7 A different equivalence

The above looks at the equivalence of the expected future cost of the loan types. It is possible to also consider the other time direction, and look at the equivalence of the expected present worth for deterministic and stochastic rates.

Consider a future amount or cash flow X_i at time i. The present worth is given by

$$PW_i = \frac{X_i}{(1+r)^i} \tag{10.18}$$

Using second order and first order approximations, respectively, based on a Taylor series, where X_i and r are uncorrelated random variables, and keeping only up to second moments, then,

$$E[PW_i] = \frac{X_i}{(1+E[r])^i} + \frac{i(i+1)X_i}{2(1+E[r])^{i+2}} Var[r] \tag{10.19}$$

$$Var[PW_i] = \left(\frac{i^2 X_i^2}{(1+E[r])^{2i+2}} \right) Var[r] \tag{10.20}$$

Equations (10.19) and (10.20) show that the expected present worth of a future sum, and the present worth variance, increase with interest rate uncertainty or variance. (This conclusion can be generalized as follows: Any investment involving net positive cash flow increases in value with rate uncertainty. And using the Carmichael equation: The value of an option also increases with rate uncertainty, a result consistent with the established options literature.)

For the expected present worth to be the same, under both fixed and variable rates, then Equation (10.18) (deterministic rate) is set equal to Equation (10.19) (stochastic rate) to give,

$$\frac{X_i}{(1+F)^i} = \frac{X_i}{(1+V)^i} + \frac{i(i+1)X_i}{2(1+V)^{i+2}} Var[r_v] \tag{10.21}$$

Simplifying gives,

$$F = \sqrt[i]{\frac{2(1+V)^{i+2}}{2(1+V)^2 + i(i+1)\mathrm{Var}[r_v]}} - 1 \qquad (10.22)$$

The above development can be extended to annual future cash flows. Consider a general investment, with possible cash flows extending over the life, n, of an investment. Let the cash flow be X_i at each time period, $i = 0$, 1, 2,..., n. The present worth, PW, is

$$PW = \sum_{i=0}^{n} PW_i = \sum_{i=0}^{n}\left[\frac{X_i}{(1+r)^i}\right]$$

where PW_i is the present worth due to X_i, $i = 0, 1, 2, ..., n$. And,

$$E[PW] = \sum_{i=0}^{n} E[PW_i]$$

$$\mathrm{Var}[PW] = \sum_{i=0}^{n} \mathrm{Var}[PW_i] + 2\sum_{i=0}^{n-1}\sum_{j=i+1}^{n} \mathrm{Cov}[PW_i, PW_j]$$

The contributions that $E[PW_i]$ and $\mathrm{Var}[PW_i]$ make to these expressions can be developed further using first order and second order approximations based on a Taylor series.

In comparison with Equation (10.17), Equation (10.22) finds that the expected variable-interest rate should always be greater than its fixed-interest rate equivalent if the present values of cash flows are to be equal. And an increase in the variance of the interest rate and the length of the loan both increase the difference between the expected variable-interest rate and the theoretically equivalent fixed-interest rate. This is generally so for all interest rates, interest rate variances and calculation period.

Exercises

10.1 Under what circumstances might the Summary results 1a, 1b, 2 and 3 (Section 10.2) not apply?

10.2 Do option types other than vanilla call options and basic expand options agree with Summary results 2 and 3 (Section 10.2)?

10.3 What degree of conservatism in investment decisions is involved in not assuming that interest rates contain uncertainty? Or is knowledge that interest rates do contain uncertainty (but not acknowledging this explicitly) used as a safety factor in case estimates of other variables are incorrect?

10.4 Does Equation (10.21) imply that as the variance of the interest rate rises, the lower the equivalent fixed rate becomes? Or is it the i'th root of something less than 1, meaning a higher equivalent fixed rate?

10.5 Using Equation (10.22), show the equivalence of fixed- and variable-interest rates on plots with axes of F and V. Within the plot, give contours of $Var[r_V]$ for different i values. Do this for different numerical assumptions.

Chapter 11

Markov chains and investment analysis

11.1 INTRODUCTION

This chapter departs from the preceding chapters on investment analysis in the presence of uncertainty, by giving an alternative method for establishing the present worth, feasibility, internal rate of return and payback period of a capital investment where the analysis variables of interest rate, cash flows and investment lifespan are uncertain. The chapter complements the other chapters in Part II. For the same assumptions, the results of this chapter's approach are the same as for that in the preceding chapters, but the approach provides additional insight into discounted cash flow (DCF) analysis under uncertainty. The method is useful for people doing investment analysis and for looking at the risks associated with investment.

The chapter uses Markov chains to model an investment. In particular, states representing different combinations of analysis variables (interest rate, cash flows and investment lifespan) and transitions between states are defined. This leads to the calculation of the probability of being in any state. With each state representing a different present worth outcome, it is then possible to calculate investment feasibility and expected present worth of the investment.

For definiteness, reference in the calculations is to present worth, but the approach is equally applicable to annual worth, future worth and benefit–cost ratio as measures of an investment's feasibility. Indicators of internal rate of return and payback period follow as a consequence.

The chapter first outlines some necessary Markov chain theory and then develops the approach through an example. It is seen that the concept of feasibility provides a unifying thread.

While Markov chains have been around for many years and have been used for a wide array of applications – marketing, finance, advertising and so on – and the literature is very large, no one appears to have looked at investment analysis in the way this chapter does.

11.2 MARKOV CHAINS

A Markov chain works in terms of discrete states and transitions between states over time. The state variables for an investment are chosen here to be the main analysis variables of interest rates, cash flows and lifespans, and together the variables define a state space (of dimension equal to the number of variables). The Markovian property implies that future behaviour depends only on the present. A Markovian assumption would appear reasonable for the present analysis; comment is given later on this. For any given investment, some states may correspond to positive present worth, while some may not.

Established Markov chain theory defines probabilities associated with state transitions, here denoted p_{ij}, in going from state i to state j; i, j = 1, 2,..., N. It follows that

$$\sum_{j=1}^{N} p_{ij} = 1 \tag{11.1}$$

and

$$0 \leq p_{ij} \leq 1 \tag{11.2}$$

Define a stochastic transition matrix **P** with components p_{ij}. For the Markov chain, the probabilities p_{ij} are a function of i and j only. The p_{ij} are taken here to be constants; comment is given later on this.

Define π_i as the probability of being in state i; i = 1, 2, ..., N, and as components of the row vector π. Then, following Howard (1960, 1971),

$$\pi = \pi \mathbf{P} \tag{11.3}$$

with

$$\sum_{i=1}^{N} \pi_i = 1 \tag{11.4}$$

Equations (11.3) and (11.4) represent N+1 equations in N unknowns, and can be used to find π_i, i = 1, 2,..., N.

Feasibility (the probability that the present worth is positive) and expected present worth are calculated from,

$$\Phi = \sum_{i \text{ with positive PW}} \pi_i \tag{11.5}$$

$$E[PW] = \sum_{i=1}^{N} PW_i \pi_i \tag{11.6}$$

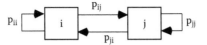

Figure 11.1 Example transition diagram. (From Carmichael, D. G., *The Engineering Economist*, 56(2), 1–17, 2011.)

If a transition diagram (for example, Figure 11.1) is used to show the states (boxes) and the transitions (arrows) between the states, then equating the inputs and outputs for each state will give Equation (11.3). In the transition diagrams given elsewhere in this chapter, transitions from a state to itself are not shown in order to not clutter the diagrams.

11.3 STATE CHOICE AND TRANSITION PROBABILITY ESTIMATION

The approach requires the analyst to establish representative states and representative estimates of transition probabilities.

Example – interest rates. For interest rates, central banks publish historical movements and there is a large amount of data that are accessible. For example, Table 11.1 shows some historical data relating to housing loan rates.

For the 120 months of data, the interest rates fluctuate between 6% and 8.5% per annum. Rounding, say, to the nearest 1% per annum interest rate, this gives 4 states – 6%, 7%, 8% and 9% per annum, where for example 7% per annum represents the interval from 6.55% to 7.45% per annum. Finer subdivision of the interest rate range will give more states, and coarser subdivision will give fewer states. The choice of states is at the discretion of the analyst. The approach remains the same irrespective of the number of states. And considering that the calculations are performed on a spreadsheet, the computational effort doesn't change with the number of states. On the matter of choosing more states, it is noted that the computations do not suffer the curse of dimensionality, but rather the number of computations is approximately proportional to the number of states.

The transition probabilities, p_{ij}, can be calculated from counting the number of monthly transitions between states and dividing by the total number of transitions, to give a frequency.

Example – cash flows. In an example multiple housing unit development scheduled over 48 months, an investor estimates that the monthly net cash flow (revenues minus costs) could vary between −$0.8M and $1.7M depending on sales and construction progress; this is separate from an initial

Table 11.1 Standard variable housing loan rate changes over a 10-year period, January year 1 – December year 10

Month	Rate (% p.a.)	Month	Rate (% p.a.)
Jan-01	6.70	Oct-04	6.30
Dec-01	6.50	Dec-04	6.05
Jul-02	6.55	May-05	6.30
Nov-02	6.80	Jun-05	6.55
Feb-03	7.30	Nov-06	6.80
Apr-03	7.55	Dec-06	7.05
May-03	7.80	Mar-08	7.30
Aug-03	8.05	May-09	7.55
Feb-04	7.55	Aug-09	7.80
Mar-04	7.30	Nov-09	8.05
Apr-04	6.80	Aug-10	8.30
Sep-04	6.55	Nov-10	8.55

capital outlay of $5.1M. In increments of $0.50M, this gives seven states: −$1.00M, −$0.50M, $0.00M, $0.50M, $1.00M, $1.50M and $2.00M. Finer subdivision of the cash flow range will give more states, and coarser subdivision will give fewer states. The approach remains the same irrespective of the number of states. The computational effort changes little with number of states. The transition probabilities, p_{ij}, might be estimated from looking at sales predictions based on the current market and economy.

Comment. These are only example ways by which states might be chosen and transition probabilities estimated. As in any discipline, estimators would combine historical data, analysis, judgement and experience to come up with what are believed to be best estimates. Interest rate estimates would be based on an understanding of likely interest rate movements in the economy, ability to borrow capital, etc. Cash flow and lifespan estimates would be based on an understanding of likely investment costs, returns and the investment environment. The estimates will also depend on the time unit or interval chosen.

11.4 EXAMPLE

To demonstrate the approach, assume a common investment profile involving an initial outlay with future returns. For definiteness, consider an initial outlay of $10 000, an annual (net positive) cash flow return of $1000, an interest rate of 5% per annum and a lifespan of 15 years.

By conventional calculations (Part I), the deterministic present worth, discounted payback period and internal rate of return for these values are 0.38×10^3, 14.2 years and 5.6% per annum respectively.

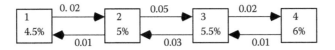

Figure 11.2 Fluctuation in interest rate; example. (From Carmichael, D. G., *The Engineering Economist*, 56(2), 1–17, 2011.)

Now allow for uncertainty in the analysis variables of interest rate, the future cash flow and the lifespan of the investment. Fluctuations could be anticipated in these variables. These variables can be considered singly and in combinations.

For fluctuations only in the interest rate, and assuming that the other analysis variables stay constant, the state is one-dimensional and the transition diagram might look something like Figure 11.2. The states are numbered in the boxes, and estimates of the transition probabilities are given next to the arrows. The actual values chosen here for the transition probabilities serve only to demonstrate the calculations.

Figure 11.2, for example purposes, anticipates that the interest rate can fluctuate plus 1% and minus 0.5% per annum about the base case of 5% per annum, while the probability of fluctuations outside this range is assumed small and is neglected.

Balancing the inputs and outputs from each state (Equation 11.3) gives

$$P = \begin{bmatrix} 0.98 & 0.02 & & \\ 0.01 & 0.94 & 0.05 & \\ & 0.03 & 0.95 & 0.02 \\ & & 0.01 & 0.99 \end{bmatrix}$$

And using Equation (11.4) and solving gives

$$\pi = [0.077\ 0.154\ 0.256\ 0.513].$$

Of the states, states 1, 2 and 3 correspond to positive present worth, and hence the feasibility of the investment is,

$$\Phi = \pi_2 + \pi_3 + \pi_4 = 0.487$$

The expected value and variance of the present worth of the investment are

$$E[PW] = 0.077 \times 0.740 + 0.154 \times 0.380 + 0.256 \times 0.038$$
$$- 0.513 \times 0.288 = -\$0.023 \times 10^3$$

$$Var[PW] = 0.077 \times 0.7632 + 0.154 \times 0.4032 + 0.256 \times 0.0612$$
$$- 0.513 \times 0.2652 = 0.035 = (\$0.186 \times 10^3)^2$$

A normal distribution can be used to model the probability distribution of present worth, however analysts can use alternative distributions if they wish. Then,

$$P[PW > 0] = 0.451$$

which is consistent, subject to the approximations made, with the earlier Φ value of 0.487.

Using the probabilities of being in any state, π_i, it is possible to calculate an expected value and variance of these state values.

$$\text{Expected value} = 0.077 \times 4.5 + 0.154 \times 5.0 + 0.256 \times 5.5 + 0.513 \times 6.0$$
$$= 5.60\% \text{ p.a.}$$

$$\text{Variance} = 0.077 \times 1.12 + 0.154 \times 0.62 + 0.256 \times 0.12 + 0.513 \times 0.42$$
$$= 0.23 = (0.48\% \text{ p.a.})^2$$

Should it be desired, then this allows a normal or other distribution to be fitted for the interest rate.

Fluctuations in the other analysis variables (cash flows and lifespans), taken one at a time, are handled similarly. Where more than one of the analysis variables (interest rate, future cash flow and lifespan) is allowed to fluctuate, then the dimension of the transition diagram would grow in proportion to the number of fluctuating variables. For example, allowing the interest rates, future cash flow and lifespan to fluctuate produces a three-dimensional transition diagram, with each dimension corresponding to one of the variables. Figure 11.3 shows an example.

Figure 11.3 anticipates that the interest rate, future cash flow and investment lifespan can all fluctuate by small amounts. As in the previous example, the probability of fluctuations outside these ranges is assumed small and is neglected. Figure 11.3 also gives estimated example probabilities associated with anticipated movements in the interest rate, future cash flow and investment lifespan. These estimates would be based on an understanding of the likely investment environment. The estimates will also depend on the time unit or interval chosen, with higher probabilities associated with larger time units.

For the Figure 11.3 case, the state probabilities π_i, i = 1, 2,..., 8 become respectively 0.076, 0.114, 0.102, 0.152, 0.095, 0.143, 0.127 and 0.190. With the last 6 states corresponding to positive present worth, the feasibility of the investment becomes,

$$\Phi = \sum_{i=3}^{8} \pi_i = 0.809$$

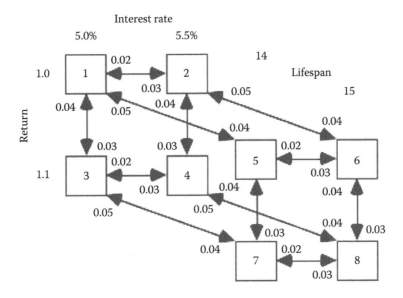

Figure 11.3 Fluctuation in interest rate, future cash flow and investment lifespan; example. (From Carmichael, D. G., *The Engineering Economist*, 56(2), 1–17, 2011.)

The expected present worth of the investment is

$$E[PW] = -0.076 \times 0.101 - 0.114 \times 0.410 + 0.102 \times 0.889$$
$$+ 0.152 \times 0.549 + 0.095 \times 0.380 + 0.143 \times 0.038$$
$$+ 0.127 \times 1.418 + 0.190 \times 1.041$$
$$= \$0.539 \times 10^3$$

11.4.1 Comparison methods

Comparison comment is given here on existing complementary methods of sensitivity analysis and the probabilistic analysis of Chapters 6 and 10.

A sensitivity analysis on the original case would vary the analysis variables of interest rate, future cash flow and investment period by plus/minus small amounts, usually one analysis variable at a time. For example, consider varying the interest rate by ±0.5% per annum either side of the assumed deterministic interest rate of 5% per annum. At 4.5%, 5% and 5.5% per annum, the deterministic present worth of the investment is respectively 0.740×10^3, 0.380×10^3 and 0.038×10^3, exactly as used in the Markov chain approach. The sensitivity analysis is showing that as interest rates increase, the feasibility of the investment decreases. The Markov chain approach additionally attaches probabilities to each different present worth calculated. In effect, the Markov chain approach incorporates the essence of a sensitivity analysis.

The probabilistic analysis of Chapters 6 and 10 uses moments of distributions to model the uncertainty in the analysis variables. For example, consider uncertainty in future cash flow, and let the annual future cash flow be described by a rectangular probability distribution with mean $1050 and standard deviation $70. States 5 and 7 of Figure 11.3, together, come closest to this. The method of Chapter 6 gives a present worth probability distribution with a mean and standard deviation (for uncorrelated future cash flows) respectively of 0.899×10^3 and 0.194×10^3, which is consistent with approximately combining state 5 (PW = 0.380×10^3) and state 7 (PW = 1.418×10^3) of the Markov chain approach. It follows that feasibility, which is defined in terms of present worth, is also consistent between the methods.

11.5 PAYBACK PERIOD AND INTERNAL RATE OF RETURN

Indicators of internal rate of return and payback period can be calculated from the above analyses.

11.5.1 Payback period

Consider first (discounted) payback period. Two-dimensional transition diagrams in terms of interest rate and return are used, such as the example transition diagram of Figure 11.4.

The lifespan is varied, and E[PW] is calculated for each lifespan to give an indication of payback period. See for example Figure 11.5, based on Figure 11.4 values. Nondiscounted payback period can be worked similarly.

As an alternative, feasibility in a payback period sense is defined as

$$\Phi = P[PBP < \text{nominated } t] \tag{11.7}$$

where t is some time. Accordingly, feasibility takes the same shape as the cumulative distribution function for PBP.

For each lifespan in a range of lifespans, E[PW] and Var[PW] are calculated, and a normal distribution is fitted to these. Since

$$P[PBP > \text{nominated } t] = P[PW < 0 \mid \text{nominated } t] \tag{11.8}$$

then the cumulative distribution function for PBP is obtained from $1 - P[PBP > t]$. Figure 11.6 shows the resulting (part) cumulative distribution function.

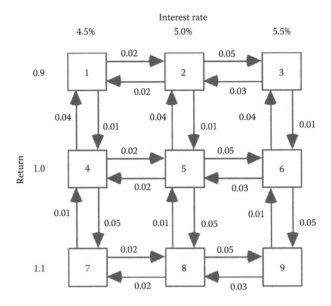

Figure 11.4 Fluctuation in interest rate and future cash flow (×10³); example. (From Carmichael, D. G., *The Engineering Economist*, 56(2), 1–17, 2011.)

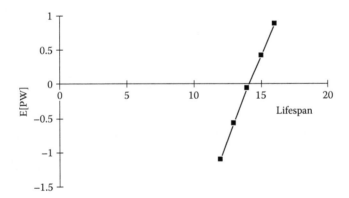

Figure 11.5 Variation in E[PW] (×10³) with lifespan. (From Carmichael, D. G., *The Engineering Economist*, 56(2), 1–17, 2011.)

11.5.2 Internal rate of return

For internal rate of return, two-dimensional transition diagrams in terms of return and lifespan are used, for example a slice through Figure 11.3 corresponding to an interest rate of 5% per annum, that is states 1, 3, 5 and 7. For a range of interest rates, E[PW] may be calculated to give an

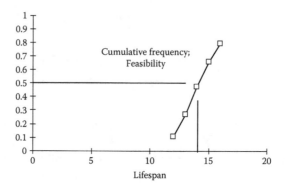

Figure 11.6 Part cumulative distribution function for PBP (years), and feasibility plot. (From Carmichael, D. G., *The Engineering Economist*, 56(2), 1–17, 2011.)

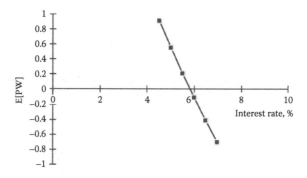

Figure 11.7 Variation in E[PW] (×10³) with interest rate. (From Carmichael, D. G., *The Engineering Economist*, 56(2), 1–17, 2011.)

indication of internal rate of return. See for example Figure 11.7, based on Figure 11.3 values.

As an alternative, feasibility in an internal rate of return sense is defined as

$$\Phi = P[IRR > \text{nominated } r] \tag{11.9}$$

where IRR is internal rate of return, and r is some interest rate. Accordingly, feasibility can be established from the cumulative distribution function for IRR, and is shown in Figure 11.8 also.

For each interest rate in a range of interest rates, E[PW] and Var[PW] are calculated, and a normal distribution is fitted to these. Then the cumulative distribution function for IRR is obtained from,

$$P[IRR < r] = P[PW < 0 \mid r] \tag{11.10}$$

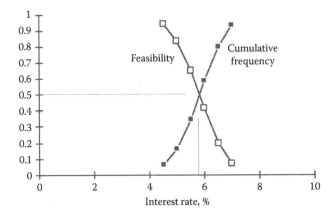

Figure 11.8 Part cumulative distribution function for IRR, and feasibility plot. (From Carmichael, D. G., *The Engineering Economist*, 56(2), 1–17, 2011.)

Figure 11.8 shows the resulting (part) cumulative distribution function.

Comment. Figures 11.5 to 11.8 are nonlinear; it is the drawing resolution which makes them appear linear.

11.6 DISCUSSION

The above exampled state transition diagrams can be enlarged by incorporating more states or finer divisions between states. The number of states and what they represent is at the discretion of the analyst. Having more states does no more than increase the amount of computations. The approach doesn't change. All the above numerical computations are readily carried out on a spreadsheet. Increasing the number of states will not make the computations any more difficult, but rather only the number of computations will increase.

On the matter of choosing more states, it is noted that the associated computations do not suffer the curse of dimensionality. The number of computations is roughly proportional to the number of states. So, for example, doubling the number of states approximately doubles the associated computations.

To keep the computations at a manageable size, some state truncation might be considered. States that are anticipated to have a low probability of occurrence could be left out. However there is no need to truncate, because the computations are readily set up and performed on a spreadsheet. A spreadsheet also enables any sensitivity to assumptions on transition probabilities to be readily examined.

The increments between states need not be constant or symmetrical as in the above examples. States are chosen to suit the particulars of the investment. In drawing transition diagrams, it is important to recognize all possible states and transitions, but having said that, there are no restrictions on the number of states or the number of transitions.

The probability of simultaneous state transitions is assumed small and is excluded. The probability of transitions outside the range of states listed in the transition diagrams is assumed small and is neglected in the above analysis, but could be included. This would lead only to a higher number of states.

In a single chapter, it is not possible to demonstrate all possible variations on transition diagrams, but those given are reasonably representative. The transition diagrams represent the investor's modelling of the investment scenario. The approach is a tool for evaluating an investment's feasibility.

Transitions between states, which are not adjacent in the above figures, may be possible in practice. For example, in Figure 11.2, it may be possible to go from an interest rate of 5% to 6% per annum in one transition. For such cases, the extra transitions, with their associated probabilities, can be inserted into the diagrams. Such cases represent no additional formulation time or computations.

State types additional to the three exampled here (interest rate, future cash flow and investment lifespan) are possible. For example, future cash flows could be broken down into lump sum and series components.

Allowance can be made for plus and minus deviations in states due to different causes, for example changes in future cash flows due to competitors or changes in future cash flows due to consumer behaviour could be introduced. This may increase the number of states and change the connectivity between states.

On an implementation issue of how to estimate transition probabilities, it is suggested that interviews, experience, and knowledge of the industry would be suitable ways to go. These estimates would be based on an understanding of likely interest rate movements in the economy, ability to borrow capital, an understanding of likely investment costs and returns and the likely investment environment. The estimates will also depend on the time unit or interval chosen, with higher probabilities associated with larger time units.

The use of Markov chains here assumes that the transition probabilities are stationary; that is the probability associated with the movement from one state to another is unchangeable. It also assumes that the chain is of first order; that is the state depends only on the immediately previous occupied state. Tests of stationarity and order have been undertaken on typical investment data, and for these data at least, stationarity and first order could be shown to be acceptable assumptions. Future research could empirically examine this assumption further.

The formulation given here uses Markov chains. That is both the state space and time are discretized. An extension to Markov processes involving continuous time could be done, as could the adoption of semi-Markov processes (where the time between transitions is a random variable) but it is not believed that the extra computation and complexity is repaid with extra knowledge about an investment.

Generally, probabilistic independence is assumed in the above analyses where needed, on the basis that any correlation information is often not available or would be hard to obtain in practice.

Exercises

11.1 Redo the calculations in the examples in this chapter. A spreadsheet should be sufficient for this purpose. Although the approach in this chapter is very straightforward, it is very easy to make a computational error.

11.2 If equations have to be set up each time an investment is analyzed, this may be off-putting to users. What is the potential for setting up a general program that can analyze any number of states containing information on interest rate, cash flow and lifespan? The user would nominate the degree of refinement or subdivision of these analysis variables. Program input would also include transition probabilities. The program would solve the relevant equations.

11.3 What is the implication of leaving out states that are anticipated to have a low probability of occurrence?

11.4 What is the implication of omitting a possible state transition, for example between nonadjacent states?

11.5 How many further dimensions beyond the three (interest rate, cash flow, lifespan) could realistically be handled by analysts?

Bibliography and references

Ang, A. H.-S. and Tang, W. H. (1975), *Probability Concepts in Engineering Planning and Design*, Vol. I, Wiley, New York.

Antill, J. M. (1970), *Civil Engineering Management*, Angus and Robertson, Sydney.

Antill, J. M. and Farmer, B. E. (1991), *Antill's Engineering Management*, 3rd ed., McGraw-Hill, Sydney.

Balatbat, M. C. A., Findlay, E. and Carmichael, D. G. (2012), Performance Risk Associated with Renewable Energy CDM Projects, *Journal of Management in Engineering*, 28(1), 51–58.

Benjamin, J. R. and Cornell, C. A. (1970), *Probability, Statistics, and Decision for Civil Engineers*, McGraw-Hill, New York.

Block, S. (2007), Are Real Options Actually Used in the Real World?, *The Engineering Economist*, 52, 255–267.

Canada, J. R. and White, J. A. (1980), *Capital Investment Decision Analysis for Management and Engineering*, Prentice-Hall, Englewood Cliffs, NJ.

Carmichael, D. G. (1981), *Structural Modelling and Optimization*, Ellis Horwood Ltd. (John Wiley and Sons), Chichester, 306 pp, ISBN 0 85312 283 0.

Carmichael, D. G. (1987), *Engineering Queues in Construction and Mining*, Ellis Horwood Ltd. (John Wiley & Sons Ltd), Chichester, 378 pp, ISBN 0 7458 0212 5.

Carmichael, D. G. (1989), *Construction Engineering Networks*, Ellis Horwood Ltd. (John Wiley and Sons Ltd), Chichester, 198 pp, ISBN 0 7458 0706 2.

Carmichael, D. G. (2000), *Contracts and International Project Management*, A A Balkema, Rotterdam, 208 pp, ISBN 90 5809 324 7/333 6.

Carmichael, D. G. (2002), *Disputes and International Projects*, A A Balkema, Rotterdam, (Swets & Zeitlinger B. V., Lisse), 435 pp, ISBN 90 5809 326 3.

Carmichael, D. G. (2004), *Project Management Framework*, A A Balkema, Rotterdam, (Swets & Zeitlinger B. V., Lisse), 284 pp, ISBN 90 5809 325 5.

Carmichael, D. G. (2006), *Project Planning, and Control*, Taylor & Francis, London, 328 pp, ISBN 13 9 78 0 415 34726 6, 10 0 415 34726 2.

Carmichael, D. G. (2011), An Alternative Approach to Capital Investment Appraisal, *The Engineering Economist*, 56(2), 1–17.

Carmichael, D. G. (2013), *Problem Solving for Engineers*, CRC Press, Taylor & Francis, London, ISBN 9781466570610, Cat K16494.

Carmichael, D. G. and Balatbat, M. C. A. (2008a), Probabilistic DCF Analysis, and Capital Budgeting and Investment: A Survey, *The Engineering Economist*, 53(1), 84–102.

Carmichael, D. G. and Balatbat, M. C. A. (2008b), The Influence of Extra Projects on Overall Investment Feasibility, *Journal of Financial Management of Property and Construction*, 13(3), 161–175.

Carmichael, D. G. and Balatbat, M. C. A. (2009), The Incorporation of Uncertainty Associated with Climate Change into Infrastructure Investment Appraisal, Conference – *Managing Projects, Programs and Ventures in Times of Uncertainty and Disruptive Change*, Sydney, unpublished.

Carmichael, D. G. and Balatbat, M. C. A. (2010), A Review and Study of Project Investment Cash Flow Correlations, *International Journal of Project Planning and Finance*, 1(1), 1–21.

Carmichael, D. G. and Balatbat, M. C. A. (2011), Risk Associated with Managed Investment Primary Production Projects, *International Journal of Project Organisation and Management*, 3(3/4), 273–289.

Carmichael, D. G. and Bustamante, B. L. (2014), Interest Rate Uncertainty and Investment Value – A Second Order Moment Approach, School of Civil and Environmental Engineering, The University of New South Wales, Sydney.

Carmichael, D. G. and Handford, L. B. (2014), Equivalent Fixed-Rate and Variable-Rate Loans, School of Civil and Environmental Engineering, The University of New South Wales, Sydney.

Carmichael, D. G., Hersh, A. M. and Parasu, P. (2011), Real Options Estimate Using Probabilistic Present Worth Analysis, *The Engineering Economist*, 56(4), 295–320.

Carmichael, D. G., Lea, K. A. and Balatbat, M. C. A. (2014), The Financial Additionality and Viability of CDM Projects Allowing for Uncertainty, School of Civil and Environmental Engineering, The University of New South Wales, Sydney.

Copeland, T. and Antikarov, V. (2001), *Real Options: A Practitioner's Guide*, Texere, New York.

Damodaran, A. (1999), *The Promise and Peril of Real Options*, Stern School of Business, New York.

Damodaran, A. (2002), *Investment Valuation: Tools and Techniques for Determining the Value of Any Asset*, 2nd ed., John Wiley & Sons, New York.

Damodaran, A. (2008), *Strategic Risk Taking*, Wharton School Publishing, Upper Saddle River, NJ.

Dandy, G. C. (1985), An Approximate Method for the Analysis of Uncertainty in Benefit–Cost Ratios, *Water Resources Research*, 21(3), 267–271.

Eschenbach, T. G., Lewis, N. A. and Hartman, J. C. (2009), Waiting Cost Models for Real Options, *The Engineering Economist*, 54(1), 1–21.

Hillier, F. S. (1963), The Derivation of Probabilistic Information for the Evaluation of Risky Investments, *Management Science*, 9(3), 443–457.

Hillier, F. S. (1965), Supplement "To The Derivation of Probabilistic Information for the Evaluation of Risky Investments", *Management Science*, 11(3), 485–487.

Hillier, F. S. (1969), *The Evaluation of Risky Interrelated Investments*, North-Holland Pub. Co., Amsterdam.

Hillier, F. S. and Heebink, D. V. (1965), Evaluating Risky Capital Investments, *California Management Review*, 8(2), 71–80.

Hodges, S. D. and Moore, P. G. (1968), The Consideration of Risk in Project Selection, *Journal of the Institute of Actuaries*, 94, 355–378.

Howard, R. A. (1960), *Dynamic Programming and Markov Processes*, Massachusetts Institute of Technology Press, Cambridge, MA.

Howard, R. A. (1971), *Dynamic Probabilistic Systems*, Volume I: Markov Models, Wiley, New York.

Huff, D. (1954), *How to Lie with Statistics*, Victor Gollancz Limited, London.

Hull, J. C. (1997), *Options, Futures, and Other Derivatives*, Prentice Hall, Upper Saddle River, NJ.

Hull, J. C. (2002), *Options, Futures and Other Derivatives*. 5th ed. 2002, Prentice Hall, Upper Saddle River, NJ.

Hull, J. C. (2006), *Options, Futures and Other Derivatives*, 6th ed., Prentice Hall, Upper Saddle River, NJ.

Johar, K., Carmichael, D. G. and Balatbat, M. C. A. (2010), A Study of Correlation Aspects in Probabilistic NPV Analysis, *The Engineering Economist*, 55(2), 181–199, 2010.

Kim, S. H. and Elsaid, H. H. (1988), Estimation of Periodic Standard Deviations under the Pert and Derivation of Probabilistic Information, *Journal of Business Finance & Accounting*, 15(4), 557–571.

Kim, S. H., Hussein, H. E. and Kim, D. J. (1999), Derivation of an Intertemporal Correlation Coefficient Model Based on Cash Flow Components and Probabilistic Evaluation of a Project's NPV, *The Engineering Economist*, 44(3), 276–294.

Kodukula, P. and Papudesu, C. (2006), *Project Valuation Using Real Options: A Practitioner's Guide*, J. Ross Publishing Inc., FL.

Lewis, N. A., Eschenbach, T. G. and Hartman, J. C. (2008), Can We Capture the Value of Option Volatility?, *The Engineering Economist*, 53(3), 230–258.

Michaelowa, A. (2005), *CDM: Current Status and Possibilities for Reform, Hamburg Institute of International Economics*, Research Paper No. 3, HWWI Research Programme on International Climate Policy, Hamburg, Germany.

NCCARF – National Climate Change Adaptation Research Facility (2009), *National Climate Change Adaptation Research Plan: Settlements and Infrastructure, Consultation Draft*, September. NCCARF, Griffith University, Brisbane.

Nielsen, L. T. (1992), *Understanding N(d1) and N(d2): Risk-Adjusted Probabilities in the Black–Scholes Model*, 16 pp., Insead, Fontainebleau, France.

Reisman, A. and Rao, A. K. (1973), *Discounted Cash Flow Analysis: Stochastic Extensions*, Monograph AIIE-EE-73-1, The American Institute of Industrial Engineers, Atlanta, GA.

Schneider, L. (2009), Assessing the Additionality of CDM Projects: Practical Experiences and Lessons Learned, *Climate Policy*, 9(3), 242–254.

Trigeorgis, L. (1996), *Real Options*, Massachusetts Institute of Technology Press, Cambridge, MA.

Trigeorgis, L. (2005), Making Use of Real Options Simple: An Overview and Applications in Flexible/Modular Decision Making, *The Engineering Economist*, 50(1), 25–53.

Tung, Y. K. (1992), Probability Distribution for Benefit/Cost Ratio and Net Benefit, *Journal of Water Resources Planning and Management*, 118(2), 133–150.

UNEP Risoe (2003), *CDM Information and Guidebook*, ed. M.-K. Lee, UNEP Risoe Centre, Roskilde, Denmark.

UNEP Risoe (2007), *Guide to Financing CDM Projects*, CD4CDM project, UNEP (United Nations Environment Program) Risoe Centre, Roskilde, Denmark.

UNFCCC (United Nations Framework Convention on Climate Change) (2011a), *Guidelines on the Assessment of Investment Analysis*, http://cdm.unfccc.int/ Reference/Guidclarif/reg/reg_guid03.pdf [Accessed 9 September 2012].

UNFCCC (United Nations Framework Convention on Climate Change) (2011b), *Tool for the Demonstration and Assessment of Additionality*, http://cdm. unfccc.int/methodologies/PAmethodologies/tools/am-tool-01-v6.0.0.pdf [Accessed 9 September 2012].

UNFCCC (United Nations Framework Convention on Climate Change) (2012), *What Is the CDM*, http://cdm.unfccc.int/about/index.html [Accessed 9 September 2012].

Wagle, B. (1967), A Statistical Analysis of Risk in Capital Investment Projects, *Operational Research Society*, 18(1), 13–33.

Index

Milton Keynes UK
Ingram Content Group UK Ltd.
UKHW040106071024
449327UK00019B/843